THE
COUNTY
FAIR
COOKBOOK

Also by Lyn Stallworth and Rod Kennedy, Jr.

The Brooklyn Cookbook

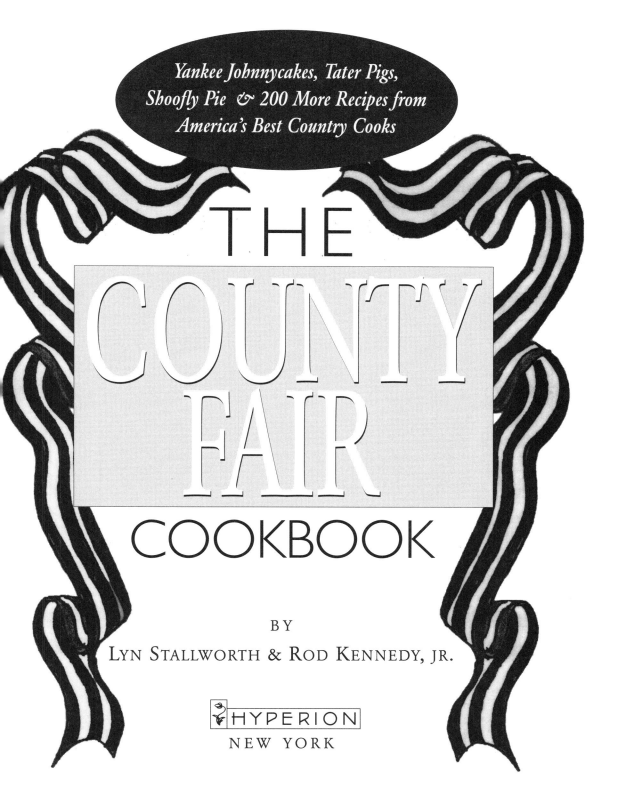

Yankee Johnnycakes, Tater Pigs,
Shoofly Pie & 200 More Recipes from
America's Best Country Cooks

THE COUNTY FAIR COOKBOOK

BY

LYN STALLWORTH & ROD KENNEDY, JR.

HYPERION
NEW YORK

Library of Congress Cataloging-in-Publication Data

Stallworth, Lyn.
    The county fair cookbook : sweepstakes ham, tater pigs, shoofly pie, and other homemade edibles / Lyn Stallworth and Rod Kennedy, Jr. — 1st ed.
        p.   cm.
    Includes index.
    ISBN 0-7868-6014-6
    1. Cookery, American. 2. Agricultural exhibitions–United States. I. Kennedy, Rod, 1944– . II. Title.
TX715.S77536 1994
641.5973–dc20   93-46035
                                                                              CIP

First Edition

10 9 8 7 6 5 4 3 2 1

DESIGN AND PRODUCTION BY ROBERT BULL DESIGN

THE SADDEST SOUND
AT THE FAIR

When the Fair is done and you're on your way
To the farm you left at break of day;
When you've started early and stayed clean through
And did all the things there was to do
When you've viewed the cattle and hogs and sheep
And was drawed into games a trifle steep
When you've tested your strength and tried your grip
Took in the sideshows and rode the "whip,"
When you've bought some peanuts without no nuts
And won a clock without no guts.
The youngsters are tired and whine and fret
With the stomach-ache from all they've et.
Your woman starts in a-jawin' you
Because you took a "swaller" or two
She's mad clear through because you went
Into the Hawaiian hula tent.
Your money's spent and you're feeling blue
For there's cows to be milked and chores to do
The hills rise up and the night comes down.
And the lights perk out in the distant town,
And up from the valley there seems to float
The saddest, weirdest, mournful note;
T'aint no use, I can't describe it half,
It's most like the blat of a dying calf,
It's the mournfullest noise I ever found
It's the last sad toot of the merry-go-round!
— MARK WHALEN
A History of the Tunbridge World's Fair

ACKNOWLEDGMENTS

Working on *The County Fair Cookbook* was, in many ways, like working on a farm.

First, surveying and clearing the land: this meant choosing the fairs for the book and dividing the country into sections.

Next, turning the soil and planting the seeds: the fairs were invited to participate, and to send us information and materials about their events.

Finally, reaping the harvest: conducting interviews and collecting the recipes and photographs accompanying the stories in the book.

Blue ribbons and our gratitude go to the hundreds of people throughout the country who helped us. We also wish to thank the thousands more we didn't speak to, whose hard work (mostly voluntary) and dedication make the county fair one of America's most popular and enduring institutions.

For extraordinary help, Best of Show rosettes go to the following: Joyce Agnew, Doug Andrews, Julie Avery, Herb Belanger, Steve Chambers, Betty Dittman, Tom Doubet, Margaret Emerson, Bob Grems, David Grimm, Beverly Gruber, June Hammond, Joyce Ice, Barbara Kirschenblatt-Gimblett, Dave Kucifer, Peggy Lipinski, Charlene Magdich, Judy Massabny, Dick Moraja, Melissa Mudd, Larry Myott, Catherine Pappas, Nancy Piche, Gay Quarty, Stacy Medlin Reed, Dorothy Reinhard, Laurette Sapah-Gulian, Donna Stollfuss, Sibyl Strates, and Jan Trojan.

Lyn Stallworth's personal thanks go to Valerie Moolman, for advice and encouragement; Jane Dystel, my excellent agent; Dorothy Drake, Ohio's best; and my wonderful testers: Sharon J. Gintzler-Gertner, Kristi Hood, Barbara Somers and Abbie Zwillinger.

Rod Kennedy's personal thanks go to Christy Archibald, Bob Bull, Earl E. Byrd, Beverly Hegmann, Edmund Kennedy, Evan Marshall, Chris Pearce, Holly Redell, and the Rev. Ron Fredrics and The Universal Spiritual Church of All Souls, for their support and guidance.

CONTENTS

INTRODUCTION

The American county fair! It's hoopla and horse pulling, kids squealing on the Ferris wheel, piglets squealing in the Swine Barn.

Across the nation, this form of bucolic blowout repeats itself yearly, mostly in late summer. It's the chance for farmers, 4-H youngsters, cooks and lots of others to compete and show their best. "But aren't all county fairs the same?" we hear you asking. The answer is yes and—emphatically—no. Yes, they all have agricultural exhibits, pie, cake, pickle and jelly competitions, boisterous midways and food gastronomically incorrect enough to give a nutritionist apoplexy. Yet each county fair is unique, a community celebration with its own distinct local essence. As you taste a sample of Creamy Rice Pudding at the Fryeburg Fair, made on a wood-burning stove from a recipe by the wood-stove cook's grandmother, you are instantly transported to a long-ago Maine farmhouse. At the Sanpete Fair in Utah, you'll hear cowgirls vying for the crown of Rodeo Sweetheart pay homage to their forebears, brave pioneering women. At the Wilson County Fair in Lebanon, Tennessee, you can tap your toes at the Fiddle Competition while munching a Goo Goo, the appropriately named marshmallow, chocolate, caramel and peanut patty beloved of country music fans and at the Yavapai County Fair in Prescott, Arizona, you'll find that Indian Fry Bread hits the spot.

The agricultural fair has been around for a long time: "You can find fairs mentioned in the Old Testament, in the Book of Ezekiel," says Lewis Miller, executive vice-president and general manager of the International Association of Fairs and Expositions in Springfield, Missouri, a leading advocate for the fair industry. "It's written, 'They of the house of Togarmah traded in thy fairs with horses and horsemen and mules.'" Historians seem to agree that a wealthy farmer and businessman named Elkanah Watson is the father of the American fair. In the early nineteenth century Watson exhibited some fine Merino sheep in Pittsfield, Massachusetts. His purpose, as behooved an idealist of the new Republic, was education; farmers could come together, see the best, and learn to improve their livestock and planting methods. "European fairs were substantially market fairs," says Julie Anne Avery, a museum curator at Michigan State University whose doctoral dissertation concerns early fairs. "Our fairs became perhaps the first adult educational activity. Elkanah worked hard to include the entire community and the family, including women's work and the arts. He encouraged women to come and engage themselves in a public event. That was a pretty revolutionary move."

"The fair also served as a means of consumer protection," says Steve Chambers, executive director of the Western Fairs Association in Sacramento, California. "Companies came to get a third-party review of products. A Santa Clara fair prize list from 1904 lists 'best set of dentures.' If you were a dentist or a guy needing false teeth, the winning manufacturer's name would mean something to you."

As rural America increasingly disappears, the fairs, paradoxically, are increasing. Fair honchos think they know why. Tom Doubet, bureau chief of county fairs, Illinois Department of Agriculture, says that city people come from a different world, and look at livestock as they would at animals in the zoo. "They'll stand in line for hours to pet a pig," says Pennsylvanian Beverly Gruber, a prominent organizer in the fair world. "People still have enough ties to rural America to pull them to the fair." Folklorist Joyce Ice, who studied New York State's Delaware County Fair for five years, agrees. "The fair's a seasonal festival, and a source of stories for the winter months. As the fair ends, everyone looks ahead to the next year's. Those who've moved away with their children to touch base with their roots, and the core values Americans don't want to lose touch with."

The Jeffersonian ideal of the happy, productive, independent rural life still holds powerful sway over the national soul. No matter that few of us will ever realize that ideal; at the county fair we join in spirit, if only for a day, with the agrarian origins of the United States.

–Lyn Stallworth and Rod Kennedy, Jr.

ONE

NEW ENGLAND

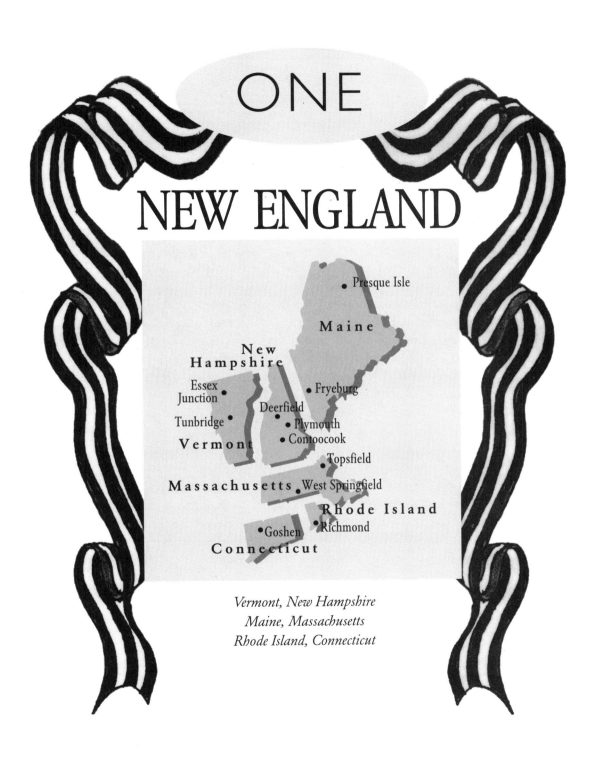

Maine

New Hampshire

Presque Isle

Essex Junction

Fryeburg

Deerfield

Tunbridge

Plymouth

Contoocook

Vermont

Topsfield

Massachusetts

West Springfield

Rhode Island

Goshen

Richmond

Connecticut

Vermont, New Hampshire
Maine, Massachusetts
Rhode Island, Connecticut

TUNBRIDGE WORLD'S FAIR

Tunbridge, Vermont

FAIR DATE: MID-SEPTEMBER, 4 DAYS

GAZETTEER: *Tunbridge is in the center of the state, between Rutland and Montpelier, the capital. Exit at Sharon from Interstate 89. The fairgrounds lie in low green meadows, in a wide bend of the First Branch River, shadowed by rugged mountains. The Green Mountains are to the west. Charming Tunbridge Village, dominated by a white-steepled church, sits above the fairgrounds. The town contains five covered bridges, through one of which visitors drive on the way to the parking area. At fair time the foliage is just beginning to turn.*

The Tunbridge World's Fair claims to trace its origin to a royal edict from King George III of England. In the original charter of the town, dated September 3, 1761, regal permission was given for the settlers to hold two fairs annually when the population had grown to fifty families. Whether or not the townsfolk took the king up on his offer is not recorded.

The grandiose title of "world's fair" was bestowed on a gathering of the Tunbridge Agricultural Society in 1867 by an orating politician, who referred to "the little world's fair." This expression met with such pleased approval that it has endured ever since.

Tunbridge Fair has ever had a reputation for conviviality. This fair serves alcohol, and one old tall tale had it that anyone found sober after 3 p.m. was fined and ordered from the grounds. Actually, a report of 1901 does state that "there was liquor on the grounds on Wednesday and several arrests were made late in the afternoon of both the hilarious and stupid." The atmosphere is much calmer now, though the beer tent does a lively business.

Tunbridge Fair's unique appeal is that a certain rural rowdiness, a serious approach to agricultural matters, and a first-rate antiquities department all coexist in harmony. Antique Hill, sited on a rise of ground, sets the tone for the fair. The quality of the collection—household furnishings, kitchen and workshop tools, old vehicles—is due to the foresight of Ed Flint, who, knowing that the technology was changing rapidly, gathered local artifacts, such as sperm oil lamps and lard squeezers, in the 1920s. The collection has grown, and is now under the care of curators W. S. Gilman and Priscilla Farnham, whose dairy farmer husband, Euclid, is president of the fair. She oversaw a newly designed colonial kitchen at the Log Cabin Museum and refurbished the turn-of-the-century post office, where you can mail letters. The general store sells old-fashioned candy, like horehound drops. Costumed volunteers man the cider press, cane chairs, demonstrate antique bicycles, stencil, dip candles and weave rugs. "Nothing here is static," says Priscilla. "What's wonderful about our exhibit is that it is living." The Ed Larkin Contra Dancers swing through the patterns of their distinctive square dances to such tunes as "Green Mountain Volunteers," played on fiddle and piano. Visitors gather round the 1881 pump organ, joining organist Katherine Whitney in singing old favorites like "Jeannie with the Light Brown Hair" and "The Old Rugged Cross." Then, if they choose, they can walk down to the midway and rejoin the late twentieth century.

The late Ed Larkin, founder of Ed Larkin's Contra Dancers.

STELLA CILLEY'S BEET RELISH

The late Mrs. Cilley was very active in the Tunbridge Fair, serving in the Ladies' Aid candy booth. She was a respected local cook. This recipe is adapted from one in the Covered Bridge Cookbook, *published in the mid-1970s by the Ladies' Aid Society of the Congregational Church. The relish is good with ham or chicken pie.*

MAKES ABOUT 2 QUARTS

4 cups finely chopped or ground boiled beets

4 cups finely chopped cabbage

½ cup grated horseradish, commercial or freshly ground and covered with white vinegar

2 cups sugar

1 teaspoon freshly ground black pepper

¼ teaspoon ground red pepper

1 tablespoon pickling salt

1½ quarts cider vinegar

1. Mix all ingredients together and pour in enough cider vinegar to just cover.
2. Refrigerate for up to 3 weeks. Or pack into sterilized jars, leaving 1/4 inch head-room, and process in a boiling water bath for 12 minutes.

ALICE FARNHAM'S MUSTARD PICKLES

A recipe from Priscilla's mother-in-law. Her husband Euclid says that his mother "most probably" exhibited them at the fair. If tiny cucumbers are not available, use all cucumber chunks. Use pickling salt, which has no additives, or kosher salt.

MAKES 9 1/2 PINTS

8 cups small whole cucumbers,
   2 to 3 inches long
2 to 3 large cucumbers, cut into
   ½-inch chunks (4 cups)
2 cups sliced onions
4 cups small whole onions, peeled
1 large cauliflower, divided into florets,
   with some of the chopped stem
3 red bell peppers, seeded and cored,
   cut into ½-inch squares

1 cup pickling salt
4 quarts water
¾ cup all-purpose flour
¼ cup dry mustard
1½ tablespoons turmeric
1 tablespoon celery salt
4 cups cider vinegar
4 cups sugar

1. In a large nonaluminum bowl, combine the cucumbers, onions, cauliflower and peppers with the salt and water. Let stand at room temperature for 24 hours. Pour off the liquid and rinse the vegetables well under cold running water. Drain thoroughly.
2. Mix the flour, mustard, turmeric and celery salt together. Add enough vinegar to make a smooth paste, then blend in the sugar. Gradually add the rest of the vinegar. In a large heavy nonaluminum pot, cook over low heat, stirring constantly, until the mixture boils and thickens slightly. Add the vegetables, bring to the boil, lower the heat and simmer for 15 minutes.
3. Pack into sterilized jars, leaving 1/4-inch headroom. Process in a boiling water bath for 10 minutes.

PARSNIP CHOWDER OR STEW

Vermonters claim that leaving parsnips in the ground over the winter makes them wonderfully sweet. As soon as the ground is soft enough in the spring, they dig them up in the short span of time before the tops grow. Serve with Montpelier crackers, split, buttered and toasted.

MAKES 6 SERVINGS

1 cup chopped onions	2 cups boiling water
2½ tablespoons rendered bacon fat	Few twists of the pepper mill
or salt pork fat	3 cups milk
3 cups thinly sliced potatoes	1 cup cream or rich milk
2 cups peeled and cubed parsnips	

1. Cook the onions in 2 tablespoons of the fat until soft but not brown.
2. Put the remaining 1/2 tablespoon fat in a heavy pot or flameproof casserole. Make thin layers of potatoes, onion and parsnips. Add the water and cook over low heat, partially covered, until the vegetables are soft, about 25 minutes. Grind on the pepper. Before serving, heat the milk and cream together and add them to the pot. The stew should have a thick, soupy consistency. Serve in soup bowls.

JESSIE BARNABY'S MINCEMEAT

For many years the late Mrs. Barnaby ran a restaurant under the grandstand during the fair. When her daughter took over the restaurant, Jessie still came, sitting in the back room. She died in 1989, at the age of 82.
Local cooks often add the optional ingredients to her mincemeat recipe. When making pies, dot the mincemeat with butter before putting on the top crust.

MAKES ABOUT 3 PINTS, OR ENOUGH FOR 3 PIES

3 cups cooked meat, preferably venison	1 to 1½ cups raisins
neck meat	5 cups sugar
6 cups chopped peeled apples	1 tablespoon ground cinnamon
1 cup apple cider	1 teaspoon freshly grated nutmeg
1 cup molasses	1 teaspoon ground cloves
1 cup cider vinegar	1 teaspoon salt
	1 teaspoon freshly ground pepper

OPTIONAL:

1 teaspoon ground ginger	1 cup strong coffee
1 teaspoon ground allspice	Grated rinds and juice of 2 oranges

1. In a large nonaluminum pan combine all the ingredients. Bring to the boil, reduce the heat, and cook slowly, uncovered, for about 2 hours or until the mincemeat is reduced to a thick consistency. The mincemeat can be frozen or canned. If canning, process the jars in a hot water bath for 45 minutes.

SWITCHEL

"Switchel is awfully thirst quenching," says Priscilla. "It's been around since colonial times, as a haymakers' beverage. At the fair, children in old-time costumes give samples to visitors."

MAKES 8 1/2 CUPS

½ teaspoon ground ginger
1 cup sugar
2 cups boiling water

½ cup molasses
½ cup cider vinegar
6 cups cold water

1. Blend the ginger and the sugar together. Pour on the boiling water and stir to dissolve. Mix in the molasses and vinegar. Let cool.
2. Add the cold water and chill thoroughly. Serve with ice cubes, if desired.

A TALE OF AN OLD HOTEL

Before campers and trailers, many fairgoers stayed at the country inns and hotels that offered fresh air to sweltering city dwellers. The Tunbridge Hotel was such a place. Fair week saw some sights: in 1897 the hotel was raided, and ninety pints of whiskey and eighteen pints of gin were confiscated. The owners were fined the curious sum of $44.67.

Another year the hotel housed two sideshow guests: a three-hundred-pound fat lady and a midget. The ladies had to keep to their rooms so that no one could see them; otherwise, why would thrifty New Englanders pay for what they'd glimpsed for free? Walking to the sideshow tent, each lady, the wide and the short, carried umbrellas draped with concealing curtains that hung to the ground.

Antique Cabin workers, 1929.

CHAMPLAIN VALLEY EXPOSITION

Essex Junction, Vermont

FAIR DATE: LATE AUGUST, EARLY SEPTEMBER, 8 DAYS

GAZETTEER: *Essex Junction is situated on the east shore of Lake Champlain, five miles east of Burlington. Nearby are the Adirondack Mountains and Fort Ticonderoga, where Ethan Allen's Green Mountain Boys soundly trounced British troops in the Revolutionary War.*

The Champlain Valley Exposition is Vermont's biggest fair, and manager David Grimm knows that successful fairs must adapt to the times: "Horse racing—harness racing—was a big part of the fair when it started in 1922, and up until the '40s. Now tractor pulls and demolition derbies draw the crowds."

Retired manager Bob Adsit, associated with fairs for over forty years, recalls that in former days "All the animals came by rail—sheep, cattle and hogs. The railroad offered exposition rates. An exhibitor could leave his farm, and be gone five or six weeks for hardly any money. As kids, we met the train and led the cattle from the rail yard to the fairgrounds. Now we have those big trailers."

Today, the entertainment in this sophisticated region of high-tech business and university life runs to big names, such as the Beach Boys and Anne Murray, but Adsit recalls simpler delights. "'A boy would race a mechanical rabbit on a one-and-a-half-mile track. The kid rarely won, but the crowd cheered him on. And the carnivals used to have wrestlers. I recall Slim Balin; he could pin them! Some fairs he made $25. Everyone who knew Slim paid a buck to go watch him pin the showman."

*Perennial prize winner
Joyce Edwards.*

SUGAR ON SNOW

Emily and Arthur (Joe) Packard, among other Chittenden County maple producers, enthusiastically volunteer at the Sugar House, built in 1975 by the Chittenden County Sugar Makers Association, of which Joe is a director. "Our purpose is to promote maple products," says Emily. "We sell maple candy, cotton candy, maple sundaes, maple coffee, and, of course, Sugar on Snow." The Packards tap around three thousand trees (some have more than one tap) and boil the sap into syrup in giant six-by-sixteen-foot cans on a blazing wood-fired special stove. "The rule is thirty-five or forty gallons of sap to make one gallon of syrup, depending on the sap's sweetness," says Emily.

Traditionally, sour pickles and doughnuts are eaten alongside this chewy treat, to cut the sweetness of the caramelized syrup. At the fair, they use Vermont Grade A Medium Amber Syrup for a robust maple flavor.

SERVES 8

1 cup maple syrup   Crushed ice

1. Bring the syrup to the boil and simmer it until it reaches 236 degrees on a candy thermometer. Divide over 8 pie plates filled with crushed ice.

GRAM'S BUTTERMILK CAKE

"I have so many ribbons I just stick them in cupboards," says Joyce Edwards, who took top honors at the Champlain Valley Fair with this cake, which she decorated with rose leaves and nuts. "I try to make things attractive," says Joyce. "I believe you eat with your eyes first." Unfrosted, this is a cholesterol-free cake. Simply dust with confectioners' sugar.

MAKES 1 CAKE

THE CAKE:

½ cup Crisco

1½ cups granulated sugar

1 teaspoon vanilla

2¼ cups all-purpose flour

1 teaspoon baking powder

¼ teaspoon ground cardamom

1 cup buttermilk

½ teaspoon baking soda

1 tablespoon water

4 large egg whites

THE MAPLE FROSTING:

½ cup pure maple syrup

8 tablespoons (1 stick) unsalted butter, at room temperature

1 pound confectioners' sugar

½ teaspoon or more Wilton's maple flavoring (optional)

- Preheat the oven to 350 degrees.
1. Grease three 9-inch round cake pans, and cut a circle of wax paper to line the bottom of each pan.
2. In a large bowl, cream the Crisco and 1 cup of the granulated sugar together until light and fluffy. Beat in the vanilla. Set the creamed mixture aside.
3. Sift the flour, baking powder and cardamom together. Set the dry ingredients aside.
4. Pour the buttermilk into a bowl or 2-cup measure. Dissolve the baking soda in the water and add to the buttermilk.
5. In a large bowl, beat the egg whites until foamy, using a stationary or hand-held electric mixer. Still beating, add the remaining 1/2 cup of sugar gradually until soft peaks form.
6. Add the dry ingredients, alternating with the buttermilk mixture, to the creamed mixture, beating well after each addition. Fold in the egg whites. Divide the batter among the 3 pans.
7. Bake for 30 minutes, or until a toothpick inserted into the center of a layer comes out clean. Cool the layers, in the pans, on racks for 10 minutes. Turn the cakes out of the pans and let cool completely on racks.
8. Make the frosting. With a mixer, cream the maple syrup and butter until fluffy and smooth. Gradually add the confectioners' sugar, beating well. Add the optional maple flavoring. Beat until the frosting is smooth and shiny, 2 to 3 minutes. Frost the 3-layer cake, and decorate as desired.

HOPKINTON STATE FAIR
Contoocook, New Hampshire
FAIR DATE: 5 DAYS, ENDING LABOR DAY

Gazetteer: Located 12 miles northwest of Concord, the state capital. Take Interstate 89, exit 6, north of 127. Fairgrounds are in the village.

Barbara Corson, a florist, has put in a quarter of a century with the fair. She is now a director, as well as superintendent of home crafts, which include knitting, sewing and food projects. "I get the judges, but I do no judging," she says. "It's not easy to get judges. I tend to go toward farm women, who know what a good thing is. One year I almost lost a judge. The problem was pickles: two jars of good-looking pickles. The judge kept looking and looking. Neither had excessive seeds, the cucumbers were sliced perfectly even, color was good, and they were firmly packed. But one jar had a bit of rust on the metal, so that one got the red ribbon. Well, that woman bugged the living daylights out of the judge. The judge said she didn't want to go through that again.

"We live in the village of Contoocook, and our house, Cranberry Barn, was built in 1790. We've heated with wood for twenty years. We have a wood-burning stove, and no one ever gets past our kitchen; they love the warmth, what with the wicked cold outside in winter. We have four chickens and a duck, and they give us enough eggs. The duck eggs are excellent for cooking.

"A few years back we decided to make maple syrup as a Girl Scout project—my husband put the gathering tank on the snow machine tractor, and I went to the maple trees up and down the street and in dooryards. The sweetest sap is in the big sugar maples with huge tops. We boiled the sap down in a pan that holds about sixty quarts. Set it on a burner outside and boiled it down to five quarts. The sap becomes maple syrup at 7 degrees above the boiling point of water. It's exhausting work. I'd go to sleep, wake up and yell, 'The sap's burning!' My husband'd tear outside in the freezing cold, come back and give me the devil 'cause I was just dreaming—everything was fine. That happened several times.

"Once we borrowed my mother's canning kettle that had been Great-grandma's to finish off the syrup. We burned a hole in it and didn't dare tell her. We were grown-ups with children, but we knew we were in trouble. My husband found another kettle like the old one, without a wooden handle. So he took a very fine saw, cut it off the old kettle. My eighty-six-year-old mother still hasn't noticed that the handle of her kettle is made of two pieces of wood, glued together!"

MY MOTHER'S DOUGHNUTS

"I grew up on the Ross Corner Dairy farm in Derry, New Hampshire," says Barbara. "My mother would get up very early and make dozens of these doughnuts to go out on the milk route, to be sold. And she sold as many as she could make. She made them with lard, not butter. These doughnuts are delicious dipped into new maple syrup."

MAKES 20 TO 28 DOUGHNUTS

2 eggs
1 cup milk
1 cup sugar
4 cups sifted all-purpose flour
4 teaspoons baking powder
½ teaspoon baking soda
1 teaspoon salt

¾ teaspoon freshly grated nutmeg
4 tablespoons (½ stick) butter, melted, or lard
½ teaspoon fresh lemon juice
Shortening or vegetable oil for deep frying
Sugar for dusting, if desired

1. In a large bowl, beat the eggs until light. Stir in the milk and sugar.
2. In another large bowl, sift the flour, baking powder, baking soda, salt and nutmeg together.
3. Stir half of the flour mixture into the egg mixture. Add the butter and lemon juice and mix until combined. Stir in the remaining flour mixture to make the dough. Cover and chill about 1 1/2 hours.
4. On a floured work surface, roll out the dough about 1/2 inch thick. Cut with a doughnut or cookie cutter into 3-inch rounds; cut out the centers and either fry them or roll again to get a larger yield.
5. In a deep heavy skillet, heat 3 inches of shortening to 365 degrees on a deep-frying thermometer. Fry 3 or 4 doughnuts at a time, turning with tongs, 2 to 3 minutes or until golden brown. Drain on paper towels. (Replace the shortening as needed.) Dust with sugar, if desired.

BARBARA'S BAKED BEANS

"Growing up on a farm, we'd already done a day's work by the time breakfast came. We always had baked beans, along with fried eggs and bacon. One year a couple of tractor puller contestants from Ohio were staying with us for the fair, and were they excited to have baked beans for breakfast—imagine, they'd never had them! They were surprised at how good they are."

SERVES 8 TO 10

4   cups dried beans: red kidney,
    yellow eye, or whatever you have
2   cups sugar, or the dark maple syrup
    made from the last sap run
1   tablespoon dry mustard

2   tablespoons molasses
½   pound salt pork
    An onion the size of a small egg, peeled
    Boiling water

1. Pick over the beans and soak overnight, covered by 2 inches of water.
2. Next day, drain the beans, cover with fresh water and parboil them for 1 hour, or until the skins split when you blow on a few beans. Preheat the oven to 300 degrees.
3. Place the beans in a 4-quart bean pot. Stir in the sugar, mustard and molasses, and bury the salt pork and onion in the beans. Cover with boiling water by 2 inches. Bake, covered, for 9 to 10 hours, topping up the pot with a little boiling water as needed.

Barbara Corson at her wood-burning stove.

PLYMOUTH STATE FAIR
Plymouth, New Hampshire

FAIR DATE: 5 DAYS, LAST WEEK IN AUGUST

GAZETTEER: *Take exit 26 off Interstate 93. Fair is located in the geographical center of the state, in the foothills of the White Mountains, surrounded by lakes; this is "Golden Pond" country.*

Pretty girl with oxen.

"Our fair is officially over a hundred and forty years old," says Fran Wendelboe, president of the Plymouth Fair, "but we are actually older. In the old days fairs rotated from town to town."

This fair has a lively present—country music concerts, monster truck shows, harness racing, chainsawing and ox pulls—and an even spicier past, with girlie shows, cancan dancers, and, not too long ago, mud wrestling. An incident reported in 1858 concerned an "eloquent" salesman and "a younger gentleman who wore a red shirt with forty-five bone buttons arranged in a diamond form on each side of the front, [and who] insinuated that the eloquent gentleman was a humbug. Eloquence therefore said Red Shirt was a 'Wolverine,' and if they had steel traps where he resided he would never have come to the fair." But fisticuffs were avoided, and the incident ended "without a sacrifice of honor by either party."

A more innocent event was the Better Babies contest: "A man came to the office not long ago with his certificate as Best Baby Boy of 1913," says Fran. "He wanted to get in touch with the Best Baby Girl of that year, to see how she'd aged ! Unfortunately, we had no means of finding out. We hadn't even known there was a baby contest."

THE VENERABLE COMPETITOR

Lora Torsey was born on a New Hampshire farm in 1896 and has been exhibiting at the Plymouth Fair for over seventy-five years. She lives with her younger sister, Minnie Smith, and four hens, two dogs and a cat. "I feed them and keep the house going, and I plant a few potatoes and a garden, and grow lots of cucumbers."

All the family loved Plymouth Fair. "One year my brother Leon Torsey, his wife Theda, cousin Lewis Gordon and wife Rosalie decided to sell honey sandwiches at the fair," Miss Torsey recalls. "Cousin Lewis took a small box with honey bees in it to call attention. They made up homemade bread, homemade butter and their honey. No one else was selling lunches, you had to bring your own, and they sold the sandwiches for twenty-five cents each. They made out very well, but never did it again."

For years Miss Torsey "took care of mothers and babies up through the Valley," and helped deliver some fifty of these babies. Her favorite meal is salt pork, "fried in the oven in an iron pan. The pork comes out lovely. I get it from my nephew; when he has a hog dressed he packs it for me. You take a big crock, and you put in an inch or so of rock salt, and you pack in strips of pork tightly, then cover with more salt. Then you add water to cover, and you put a plate over it and a rock to hold it down. You put it in the cellar and it will last a year."

LORA TORSEY'S GOLDEN GLOW

"Use cukes as they come from the garden," says Miss Torsey. "Never soak them in water as they take up some."

MAKES APPROXIMATELY 5 QUARTS

4 quarts cucumbers, washed and sliced

2 medium green bell peppers, cored, membranes removed, cut into small pieces

2 medium red bell peppers, cored, membranes removed, cut into small pieces

4 medium onions, peeled and sliced

½ cup pickling salt

1 quart cider vinegar

4 cups sugar

Spice bag:

3 tablespoons whole mixed pickling spices

1 teaspoon celery seed

1 teaspoon mustard seeds

1. Place the cucumbers, peppers and onions in a nonaluminum bowl and toss with the salt. Leave overnight.
2. In the morning, drain the mixture in a colander and rinse well under running water.
3. In a large nonaluminum kettle, bring the vinegar, sugar and all the spices, tied in a bag, to the boil; stir so that the sugar dissolves. Add the vegetables. Return the mixture to the boil, then lower the heat to a simmer for twenty minutes. Remove the spice bag and pack at once into hot sterilized jars, leaving 1/4 inch head space. Adjust caps and process 15 minutes submerged in a boiling water bath.

YANKEE JOHNNYCAKE

"This is a dish my mother made way back, to eat with baked beans on Saturday night," says
Lora Torsey. *"It always comes out great. I used to put in a handful
of pork scraps when we had them, and that was wonderful. The batter will be thin. When
my father went down to Mississippi one winter, I made him one
and sent it by mail."*

MAKES 1 LARGE CAKE

1 egg	2 teaspoons baking soda
⅔ cup molasses	1 cup all-purpose flour
3 cups buttermilk	2⅔ cups sifted cornmeal
1 teaspoon salt	

- Preheat the oven to 325 degrees.
1. Grease a large pan, such as a 13 × 9 × 2-inch Pyrex baking dish.
2. In a large bowl, beat the egg. Add the molasses and buttermilk and stir in the salt, soda, flour and cornmeal. Pour the batter into the dish.
3. Bake 35 minutes, or until the cake is set. Cool slightly on a rack; serve warm.

DEERFIELD FAIR
Deerfield, New Hampshire
FAIR DATE: 4 days, first week in October

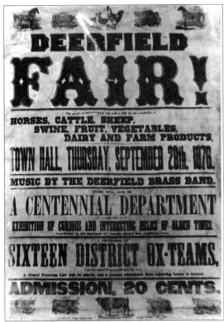

1896 poster.

GAZETTEER: *Equidistant from Concord and Montpelier, off Route 107, outside the village of Deerfield.*

Deerfield held its first fair in the center of town in 1876, the country's centennial year. It was the biggest event of the local season, and it still is. For a century—up until 1976—the fair began with a parade, complete with marching bands, patriotic floats, decorated wagons carrying vegetables and pretty girls, and wagons drawn by nineteen yoke of oxen, like one entry of 1878. The parade was, alas, a victim of popularity; now there is just too much traffic, too many people.

In 1938 the fair moved out of town to its present site, a former CCC camp built under Franklin Roosevelt's New Deal. Although greatly expanded to offer events from thirty-six-hundred-pound oxen pulls to features like the Granite State Cloggers, "The Family Fair" remains Deerfield's motto, and organizers zealously see that it stays that way. In the 1940s the Reverend Carolyn Scott found, to her horror, that a striptease show was in rehearsal. "Walter!" she shouted to her husband, the fair director, "there's a woman with no clothes on over here; get her off the fairgrounds!" Naked performers are still banned. Deerfield is proud to be an old-fashioned agricultural fair with enduring values.

The Pig Scramble.

TWO RECIPES

from the *Deerfield Fair Cookbook*, 1911

PICKLE FOR BEEF

*Old-time recipes were short on directions; it was assumed that you knew how to cook.
However, should you want to make a vast quantity of corned beef, here are the proportions.
Good luck!*

MAKES LOTS!

100 pounds meat	3 pounds brown sugar
4 gallons water	1 ounce saltpeter
8 pounds salt	1 teaspoon soda

Sprinkle a little of the salt between each layer of meat. Pour the brine over meat scalding hot.

MARY ANNS

*Abbie Zwillinger, who tested this recipe, suggests adding 1 teaspoon of any flavoring
to the batter: vanilla, almond, lemon, or orange. Drop from a teaspoon on a buttered tin.
Bake in a quick oven. Watch closely, they burn quickly.*

MAKES 3 DOZEN 2 1/2- TO 3-INCH COOKIES

2 beaten eggs	¾ cup flour
1 cup brown sugar	½ teaspoon baking powder
	1 cup walnuts cut in small pieces
A pinch of salt	

Preheat the oven before you begin. The "quick oven" is 375 degrees.

Combine the batter quickly, and drop spoonfuls two inches apart onto baking sheets. Bake 4 to 5 minutes; cool on racks.

Almost…not quite.

NORTHERN MAINE FAIR

Presque Isle, Maine

FAIR DATE: First full week in August

GAZETTEER: *The fair is held on Mechanic Street in Presque Isle, Aroostook County, Maine's largest county, close to the Canadian border. The area's attractions are its lakes in summer, moose and deer hunting and skimobiling in winter.*

Aroostook County 1839-1989

Presque Isle Park, July 28, 1889

Northern Maine Fair

135th Consecutive Year
Presque Isle
August 5 - 13, 1989
List of Premiums

"The emphasis is on young people at our fair," says Gaylen Flewelling, past president of the fair. "There is no other fair in this area, and we get mostly local people. High school graduates, wherever they are, come back during fair week to see their friends." The fair features over thirty kinds of potatoes, all exhibited by high school students.

As president, Gaylen Flewelling oversaw the fair's organization and smooth running. He and his wife Joan are potato farmers. "I'm the fifth generation in this business," he says. It's fitting that the fair honors young people, because they play an important part in the potato harvest. Maine is the only state that recesses school in mid-September for three weeks (or longer if the weather slows the harvest) so that kids can work in the fields. "We have a hand-picking crew of about twenty-five," says Gaylen, "and they must be twelve years old. At the age of sixteen, they can go on the harvesters." The little ones earn a hundred dollars or so each week, and the older ones make two hundred. "The child labor laws are strict," says Joan. "They must have permission from their parents, and a certificate of age from the school. We have a good work ethic here. In our town of Easton, 90 percent or more of the children work. They may blow money they get from their parents, but the potato money goes for things like contact lenses, boots, or winter coats."

The Flewellings raise potatoes for a tough customer; Wendy's fast food chain buys much of their crop for french fries. "They're fussy, I can tell you that!" says Gaylen. "Potato raising is a matter of hard work and luck," he notes. "We don't have sunshine every day, and we don't irrigate."

"Each potato is different," Joan adds. "Some go for potato chips, some for fresh pack. Round whites are wet and russets are dry; russets are best for potato salad."

Surrounded by spuds as they are, the Flewelling family genuinely likes Maine potatoes. Besides the familiar baked and boiled everyday side dishes, Joan has found a way to use them in a delicious candy.

JOAN FLEWELLING'S POTATO CANDY

Joan's mother, Patience Clark, got this recipe years ago. The candy is a fluffy fondant.
Mound it on a plate and it tastes very much like a well-known candy bar.

MAKES ABOUT 32 (1-INCH) SQUARES

½ cup freshly mashed potato

3 cups sifted confectioners' sugar

1 cup sweetened or unsweetened
   shredded coconut

1 teaspoon vanilla extract

2 ounces sweet or semisweet chocolate

1. Butter a 7-inch pie plate or similar dish. Place the mashed potato in a bowl and add the sugar, coconut and vanilla. Stir with a wooden spoon until well mixed.

2. Transfer the mixture to the buttered pan. Melt the chocolate and pour it over the top of the candy, spreading it with a butter spatula. Place the pan in the refrigerator for 30 minutes. Before the candy sets and the chocolate hardens, score to make squares. Let the candy set in the refrigerator, then cut through the squares. Kept covered in the refrigerator, the candy will last 10 days—if you can keep it that long.

The Flewelllings and Heidi in Superior potato field, Culberson Farm.

FRYEBURG FAIR

Fryeburg, Maine

FAIR DATE: ALWAYS THE FIRST WEEK IN OCTOBER CONTAINING THE FIRST WEDNESDAY IN OCTOBER, FOR 8 DAYS

GAZETTEER: *Fairgrounds located on Route 5 outside Fryeburg in western Maine, very near the New Hampshire border and scenic White Mountains. Fertile farmland, spectacular fall foliage.*

"An Old Time Country Fair, Enriched with the Passage of Time." "That's the fair motto," says Philip G. Andrews, a dairy farmer who is the fair's president. "It's the last fair of the season in Maine, and pretty much the last fair in New England. It's the end of the summer's competition, and we get more livestock than other fairs that I know, over three thousand head. People come and stay all week. We have the largest camper hookup for electricity and water outside of Disney World!

"I first attended when I was about four years old, and I started on the finance committee in '48. We're a fair family. My oldest son heads security, and another son is general superintendent. All my children and grandchildren have lifetime memberships, and I'm now buying them for my great-grandsons."

Loretta Greene's kitchen is perhaps the biggest draw at the Fryeburg Fair Farm Museum, an exhibit of old-time farm crafts. After all, when you get the chance to watch an expert cooking in and on a wood stove just as your great-grandmother did, it's interesting. And when you get to sample the food that comes from that stove, the experience is irresistible.

"I gave out seven thousand two hundred and forty-odd samples this last time," says Loretta, "of beans, rice pudding and apple crisp. And that doesn't count the peanut brittle, biscuits and johnnycakes that they just pick up and walk away with." (Loretta gets the total of what she gives away by counting the little serving containers.)

The entire Greene family shows visitors what farm life was like in the old days. Ted, Loretta's husband, is a maple syrup producer, and son Alan gives a syrup-making demonstration. Ted also demonstrates the old craft of shingle-making, and put a lot of work into building the windmill, showing the crowds how it was used to pump water. Son Jason attached an antique corn grinder to what he calls a "one lunger" engine and ground both coarse and fine cornmeal. Daughter Tara's job is peeling apples and washing dishes.

"About four years ago I saw the wood stove that someone had donated to the museum," Loretta recalls. "It was hooked up to a chimney, but stone cold and dirty. 'Too bad it's not used,' I said. And they said, 'Go to it! We'll buy the ingredients.'"

Loretta knows what she's doing because she cooks on a wood stove at home almost every day of the year. "About eighteen years ago, back when oil prices went up, my husband remembered a wood stove he'd got at an auction of his great-aunt's estate long ago. It was in the barn. We built a chimney and brought it in. For the first six months I kept my regular gas stove, but I found out how enjoyable and easy the wood stove is. You can keep a wood stove at 350 degrees just like a regular stove; you just have to know your wood. If I had my choice, I'd always burn ash. It's clean, and it splits neat, not a lot of splinters. It gives a nice even heat, and lasts a good while. If you want to slow a fire down, throw in a piece of oak. Oak over ash will last for some time. On a cold night, to keep the stove going overnight, we sometimes add a chunk of hard maple or apple with oak.

"The best part of doing the cooking at the fair is the chance to show folks the practicality and enjoyment of wood stove cooking—a lot of people have wood cook-stoves, and most don't use them, except maybe to put a teakettle on. They've never dared to do anything in the oven. I get a lot of questions about cleaning the stove, finding parts for it, and regulating the oven for baking."

Loretta and Ted both work outside their farm. She leaves at 4:45 a.m. for her job at the L. L. Bean company, in Freeport. "I'm home early enough to enjoy the evening," she says. And to cook great food.

Animal oddities.

MY GRANDMOTHER'S CREAMY RICE PUDDING

Rose Robinson, Loretta's grandmother, lived on a small farm in South Paris, Maine. "This is the version as it's been passed down," says Loretta.

SERVES 4

1 cup white rice
1 cup cold water
4 cups whole milk
½ cup sugar

½ teaspoon salt (optional)
½ cup raisins (optional)
Few drops lemon extract

1. Boil the rice and water until the water is absorbed.
2. Add the milk, sugar and salt, and the optional raisins. Simmer uncovered over low heat (or on the back of a wood-burning stove) until most of the milk is absorbed and the rice is tender. Stir in the lemon extract.

Loretta Greene at the wood-burning stove.

INDIAN PUDDING

"Delicious with whipped cream," says Loretta.

SERVES 4 TO 6

6 cups whole milk
½ cup yellow cornmeal

⅔ cup molasses
1 teaspoon salt

• Preheat the oven to 325 degrees.
1. Heat 4 cups of the milk until scalding, or just until little bubbles begin to form around the edge of the pan. (Refrigerate the remaining 2 cups until needed.) Add the cornmeal slowly, stirring constantly. Keep stirring until the mixture thickens. Stir in the molasses and salt; remove the pot from the heat.
2. Pour into a 9 × 13 × 2-inch Pyrex baking or earthenware dish. Pour on the cold milk; no need to stir it in. Bake for 3 hours.

TOPSFIELD FAIR

Topsfield, Massachusetts

FAIR DATE: EARLY OCTOBER, 10 DAYS

GAZETTEER: *The town of Topsfield is 30 miles northeast of Boston, exit 50 off Route 95. The fairground is on Route 1 in Topsfield. The terrain of rolling hills is 15 miles from the seacoast, the resort area of the North Shore, popular with Bostonians.*

"We claim to be the oldest *continuous* fair in the United States," says Caroline Craig, whose husband, Al, is fair manager. "The fair was first held in 1818 under the Essex Agricultural Society. The only hiatus in all our history was two years during the Second World War." The surrounding area is farmed less and less, and "We get agriculture for the fair wherever we can find it," says Caroline. "Putting on a fair, you have to worry about everything from major entertainment to who cleans the restrooms."

The Craigs are caretakers at a historic property maintained as a working farm, owned by the Society for the Preservation of New England Antiquities, whose goal is to maintain it as a living museum. The property is venerable; J. Cogswell received a deed to the property from King George II in 1735.

For the fair, Caroline is in charge of the draft horse competition, begun in 1980, to replace greyhound racing. "The betting was getting out of hand, and the facility was dilapidated," says Caroline. (The Craigs own seven big Belgian horses.) As farm horses are no longer needed for plowing and other heavy work, they've become a thriving hobby, and have moved into the show ring. Performance classes include single-horse hitch, six-horse hitch, and classes simulating farm work. Four breeds are shown: Percherons, Shires, Belgians and Clydesdales. "These competitions are a big deal," says Caroline. "Some of the major exhibitors go from coast to coast."

Raised on a "typical small New England farm" in New Hampshire, Caroline "loved everything but the weeding, and I still do."

Team driving competition.

CAROLINE'S CRANBERRY CHUTNEY

Make the chutney in fall when cranberries are in season. Packed into jars, it keeps well in the refrigerator for up to two months. Or process the chutney in sterilized jars in a hot water bath.

MAKES 8 TO 10 CUPS

2 whole navel oranges, cut into chunks	¼ cup distilled white vinegar
6 cups fresh cranberries	1 tablespoon finely chopped fresh ginger
8 ounces golden raisins	2 to 3 cups sliced almonds
4 cups sugar	

1. In a food processor, chop the oranges and cranberries. Do not chop too fine: you want some texture.
2. In a large nonaluminum kettle or soup pot combine the orange-cranberry mixture, the raisins, sugar, vinegar and ginger. Bring to the boil over moderate heat, stirring frequently. Boil gently, until the liquid has reduced to a syrupy consistency, about 40 to 45 minutes. Remove from the heat and add the almonds. Pack into jars.
3. Refrigerate or process in a boiling water bath for 20 minutes.

HURRY-UP BROCCOLI CASSEROLE

"You can serve this as a main dish or a side dish," says Caroline. "For company, I serve it surrounded by puréed winter squash."

SERVES 4 TO 6

2 (10 ounce) packages frozen cut-up broccoli, thawed, or the equivalent	3 eggs, lightly beaten
	3 tablespoons flour
8 ounces cottage cheese	1 teaspoon salt
6 ounces cheddar cheese, grated (1 ½ cups)	1 teaspoon dry mustard
	¼ teaspoon freshly ground pepper

• Preheat the oven to 350 degrees.
1. Mix all ingredients together, only enough to blend. Spoon into a greased 8 × 8-inch Pyrex baking dish or equivalent casserole.
2. Bake until bubbling and browned, approximately 30 minutes.

ANADAMA BREAD

Instead of hand-kneading, you can make this bread in a standing heavy-duty mixer equipped with a dough hook, kneading it in two or three batches. "I deliberately underbake my bread," says Caroline Craig. "Overbaking is the biggest mistake a home baker can make. It kills the flavor."

MAKES 4 TO 6 LOAVES

3 cups yellow cornmeal

6 cups water

12 tablespoons (1½ sticks) unsalted
    butter, cut into pieces

1½ cups molasses

2 tablespoons plus 2 teaspoons salt

3 packages active dry yeast

6 eggs

10 to 12 cups all-purpose flour

1. Stir the cornmeal into the water in a large saucepan and bring to the boil, stirring constantly. Remove from the heat and add the butter, molasses and salt. Let the mixture cool to lukewarm (110 to 115 degrees). Sprinkle the yeast over the mixture and let soften for a few minutes.

2. Transfer the batter to a large mixing bowl. Stir in the eggs.

3. Add 6 cups of the flour gradually, stirring to make a stiff dough. Turn out on a floured board and begin to knead; add more flour as needed to make the dough manageable. You will add 4 to 6 cups more, depending on the humidity. Knead 10 to 12 minutes, until the dough is smooth and elastic.

4. Put the dough in a large greased bowl with plenty of room for the dough to expand (a soup kettle or other large cooking pot will do just as well). Cover with oiled plastic wrap or a lid and let rise until it doubles, about 1 1/2 hours.

5. Form the dough into loaves and fit into oiled bread pans, to make 4 large or 6 small loaves (use 4 × 8 × 3-inch pans, or 3 1/2 × 7 1/2 × 2-inch pans). Cover lightly with kitchen towels and let rise again until dough is doubled, about 1 hour. Meantime, preheat the oven to 375 degrees.

6. Bake the loaves until they begin to brown, about 10 minutes. Reduce the heat to 325 degrees and continue baking. Small loaves will need about 20 additional minutes, large loaves 30. Remove from the pans and cool on racks.

THE MOTHER OF THE MIDWAY

The agricultural displays are the core of any fair; the sleek, fleshy milk cows, the well-groomed goats, the earnest 4-H'ers' entries, the produce, the jams, and the pickles. But a fair is incomplete without its friskier side—the midway! Games of chance and skill; thrilling rides; irresistible, indigestible, calorically colossal treats like fried dough; and over all, the oom-pah beat of the music that quickens the blood and cheers the heart.

Fiesta Shows, "A Family Entertainment Tradition Since 1925," runs three units of games, rides and food facilities serving fairs throughout New England. For big fairs, such as Topsfield, the units come together. As many as sixty people tend the rides, running them, putting them up and dismantling them. Another sixty or eighty work the games. And they all need to eat.

Connie Marra sees that they do. For her Fiesta unit, she runs the cookhouse. "Each show that goes out on the road in spring has its own cookhouse," says Connie, "and it's the last place to close when the show is over. On a Sunday night I'll stay until the last piece has left the lot. Then we go on the road at three or four in the morning, sleep in the house trailer or truck we've arrived in for a few hours, and the cookhouse is the first to go on location at the new spot. We're the first because everyone has to eat before setting up the rides. The boys set up the water lines and the electricity, and I'm in business."

By the time breakfast is over (bacon, eggs, french toast, fifteen kinds of cold cereal), Connie and her staff start preparing what will be lunch or dinner, depending on a worker's breaks.

"From one or two p.m. until around nine or ten p.m. we serve about five types of meals," Connie says. "There's always a pasta, steak and cheese and meatball subs, and a chicken and a pork meal. A person may have a supper break at two p.m. or at 7 p.m. We open at six in the morning, and stay open for an hour after the fair closes.

"The cookhouse is also a mobile medical center for the lot. There's ice, bandages, aspirin, help and a kind word. One of our bosses calls me 'Mother of the Midway.' I collect when there's a death or someone suffers a misfortune. We have a 'swear jar' in the cookhouse, and anyone who dares to curse pays up. The money goes to the scholarship fund for people in the business; contrary to what some think, there are a lot of educated people in the carnival. I keep my eye on them all, and lend a few bucks here and there. I must say I've always got my money back. They're very respectful boys for the most part. We go out for shows around the first of March, and stay out until November first. I don't think I'd enjoy any other kind of work."

CARNIE CHICKEN ITALIANO

Carnival cook Connie Marra's recipe calls for 10 1/2 pounds of chicken.
We've reduced it to a more manageable size.

SERVES 6

1 (1-pound 4-ounce) can crushed tomatoes

1 (14-ounce) jar commercial spaghetti sauce

⅓ cup olive or vegetable oil

1 tablespoon crumbled dry oregano

Salt and pepper to taste

3½ pounds chicken parts

1½ large onions, peeled and sliced lengthwise

1½ large green peppers, cored and sliced lengthwise

2½ pounds white potatoes, peeled, each sliced lengthwise in 8 pieces

- Preheat the oven to 325 degrees.
1. In a large baking dish, combine the crushed tomatoes, spaghetti sauce, oil, oregano, salt and pepper. Mix well. Add the chicken in one layer. Top with the onions, peppers and potatoes, and add water just to cover.
2. Cover with a lid or seal the pan with aluminum foil. Bake for 45 minutes. Remove the lid or foil and cook 45 minutes longer.

Wedding on the Giant Gondola Wheel: concession worker Denise Ahearn and ride operator Louis Tremblay, with attendants.

The Midway!

GAZETTEER: *Near Richmond, Rhode Island, on Route 112, about 1 mile south of Route 138, heading east to the yachting and tourist town of Newport.*

The impetus behind the fair was the Grange, a nationwide fraternal and agricultural society. Eleven local units of the Grange bought a tract of land to establish a fairground, now owned and operated by the Washington County Pomona Grange. The fair is a means to tell the world about the Grange and its good works, especially with the deaf.

Virginia Cottrell and her late husband Jay were stalwarts of the Richmond Grange and were among the organizers of the Washington County Fair in 1967.

"We at the Grange decided we wanted a family-oriented fair, and not a carnival atmosphere," says Virginia. "When we started, the men cut the posts, the women peeled off the bark, and we put up the buildings. The fair has grown by leaps and bounds—lots of trees, a real rural atmosphere, and we think it's beautiful! Almost all the labor is free, except for car parkers and security. We give our time all year long for the fair. Our son Jack works on the entertainment committee.

"My husband was a carpenter, and we also went into raising chickens for the kosher Jewish trade in New York. We started with a thousand chicks. Raising them is a science. You must keep them very clean, because the ammonia smell gives them a cough. And if someone blows a horn in the middle of the night, forget it. They're very skittery as well as delicate. Middlemen came and caught the chickens, and took them, alive, to New York.

"You have to grow the chicken so that between its legs you can feel a blob of fat. Lots of good yellow fat means a good price. But the market prices could fluctuate. We weren't idle a minute. My husband would do his chores, work eight hours in his carpentry shop, come home, have supper, do more chores, and then we'd work on the fair, or go Granging.

"We'd visit the various Granges just to socialize. Grange members are called Patrons of Husbandry, but we all changed that 'P of H' to 'pack of hogs.' Grangers are known for eating, and you can expect really good refreshments."

COUNTRY POTATO SOUP

You can, of course, use chicken broth instead of water and bouillon cubes. "Makes a hearty lunch, with crackers," says Virginia.

SERVES 4

3 cups diced potatoes
½ cup diced celery
½ cup diced onion
1½ cups water
2 small chicken bouillon cubes or 1 large

½ teaspoon salt
2 cups milk
2 tablespoons butter
1 cup sour cream

1. In a large pot, combine the potatoes, celery, onions and water. Add the bouillon cubes and salt. Bring to the boil, lower the heat and simmer, covered, for about 20 minutes, until the potatoes are tender.
2. With a potato masher, mash the potatoes slightly, so that they retain a little texture. Add the milk and butter. Heat, but do not let boil. When ready to serve, stir in the sour cream.

GOSHEN FAIR
Goshen, Connecticut

FAIR DATE: Early September, 3 days

GAZETTEER: *Route 63, south of Goshen Center. Goshen is in the northwest corner of Connecticut, near the almost ideal New England village of Litchfield, birthplace of Green Mountain boy Ethan Allen, the Revolutionary War hero. Litchfield and the surrounding towns and countryside of Litchfield County are well worth seeing; this is picture postcard New England, and also prime antiquing country.*

**GOSHEN FAIR
LABOR DAY WEEKEND**

Sept. 5, 6, & 7, 1992

**Goshen
Fair Grounds**

Route 63, Goshen, CT

"This was dairy farming country," says Richard Kobylenski, president of the fair, "because up here grass and corn for feed grow well. Around Goshen when I was a boy there were eighty dairy farms, and now only two are active. But we still put on a terrific dairy show; we had two hundred and thirty head of cattle this year. My dad was a farmer; he came from Poland. When he came here, it was all Yankees and Irish. Now there's a mix of Slovaks, Italians, Irish and Polish. And everybody loves the Goshen Fair!"

"We place our posters all over the countryside to advertise the fair," says Jean Breakell, a board member who is in charge of publicity. "We began with a cow, and each year there is a different animal. Virginia Anstett is the artist. We're happy to sell posters, too; some people have the whole series, framed and hanging in their houses. So far, the draft horse has been most popular."

Fair secretary Katherine Vaill won almost more blue ribbons than she could hold in 1992—for raised rolls, Parker house rolls, raised currant cake and cinnamon rolls. "I was just lucky!" she disclaims modestly. "I really like to work with yeast."

Katherine has lived on a farm all her life—first as a child in Harwinton, Connecticut, then, after she married, in Goshen, on one of the remaining dairy farms. "My husband passed on and my son Donald farms now. He milks around sixty cows; it always was a Guernsey herd, but now there are some Holsteins mixed in," she says. As secretary, Katherine receives entries, makes out checks for premiums, and attends meetings year round. "It's a big part of my social life," she says.

RAISED CURRANT CAKE

"To this day I don't remember where I got the recipe," Katherine Vaill says. It makes
a beautiful, impressive cake, good with tea or coffee, or anytime. For ease and best results,
use a standing electric mixer.

MAKES 1 LARGE CAKE

FOR THE SPONGE:

2 cups all-purpose flour
1 teaspoon salt
1 cup milk, scalded and cooled
    to lukewarm

½ cup warm water
1 package active dry yeast
4 teaspoons sugar

FOR THE CAKE:

2 cups granulated sugar
2 sticks unsalted butter, at room
    temperature
2 large eggs
4 teaspoons milk
2 teaspoons vanilla extract

2 cups sifted all-purpose flour
2 teaspoons baking powder
2 teaspoons freshly grated nutmeg
¾ cup currants
    Confectioners' sugar

- Preheat the oven to 350 degrees.
1. Make the sponge. In a bowl combine the sponge ingredients, mix well, cover and let stand for 30 minutes, or until bubbly.
2. Grease a 10-inch Bundt pan, preferably nonstick. Meanwhile, using an electric mixer at slow speed, cream the sugar and butter until fluffy. Add the eggs and beat well. Add the milk and vanilla and beat to incorporate. Beat in the flour, baking powder and nutmeg.
3. Add the sponge, beating only enough to incorporate, then beat in the currants. Pour into the Bundt pan, smooth the top, and bake in the center of the oven for 50 to 60 minutes, or until a tester inserted in the middle comes out clean. Cool on a rack for 10 minutes. Turn out of the pan and return to the rack. When completely cool, dust with confectioners' sugar.

EASTERN STATES EXPOSITION—"The Big E"

West Springfield, Massachusetts

FAIR DATE: MID-SEPTEMBER, 12 DAYS

GAZETTEER: *The Big E, officially the Eastern States Exposition, 1305 Memorial Avenue, West Springfield, Massachusetts.*

Make no mistake, the Big E is BIG. Over one million fairgoers each year enjoy what is in essence New England's own fair. There is something for every one of those visitors. Attractions range from the homespun—the First Congregational Church's concession, serving family-style turkey dinners with all the fixins, including butterscotch pudding and gallons of clam chowder—to the elegant—the white-tablecloth Storrowton Tavern Restaurant. There's an A-rated horse show; a midway with a giant, stomach-lifting slide; and Storrowton, a reconstructed nineteenth-century New England village for the historically minded.

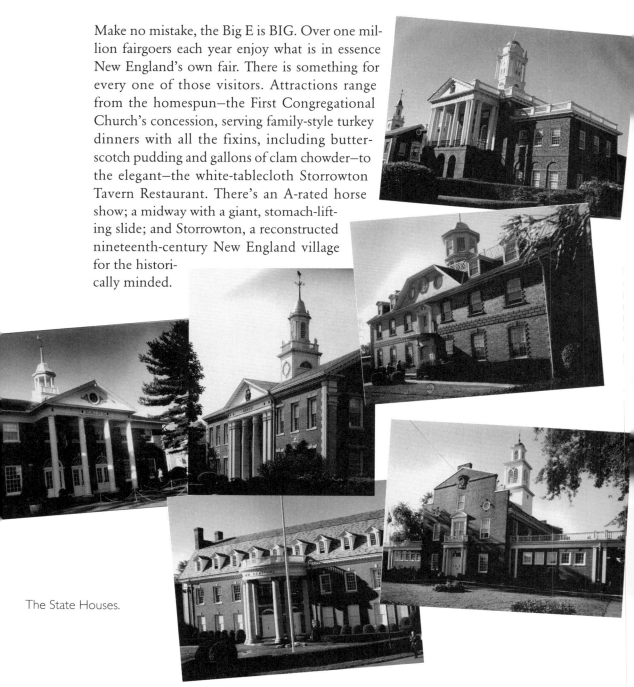

The State Houses.

The pride of the Big E, however, is its Avenue of States: six buildings representing the capitols of Massachusetts, Vermont, New Hampshire, Rhode Island, Maine and Connecticut. Each state owns its piece of land and its building, and

COUNT YOUR CHICKENS WHILE THEY HATCH

Thousands of people eagerly cross the road to the Big E's Farm-A-Rama building to watch baby chicks peck their way out of the shell. Around twenty-two hundred chicks hatch, in relays, during the fair's twelve-day run. Harry Adriance, the operator, designed and built the twelve-by-eight-foot incubator. Everything possible is done to coddle these future chickens. Fans and circulating air keep the hatchery at an even 100 degrees, and a humidifier keeps the air moist. "Otherwise," says Adriance, "the shells get hard, and the chicks can't peck their way out of hard shells."

Eggs take twenty-one days to hatch, and Adriance must pace his fertile eggs so that a number of chicks hatch each day. He starts them in conventional incubators at his farm, and two days before the fair he loads them into his van and drives v-e-r-y carefully to the fairgrounds. He stays near the hatchery, which is crowned by a giant plastic chicken, throughout the fair. (Incubating eggs must be turned twice a day.) The chicks are donated to the Future Farmers of America for vocational projects. "Fair organizers always say they want an incubator like this," says Adriance. "When they find out how much work it is, we never hear another word."

Chick hatchlings and spectators.

uses the property to display its products and regional enterprises. You can sample seafood in the Rhode Island building; find maple sugar candy in Vermont's; and join the long line for baked potatoes with myriad toppings in the Maine building, a replica of Maine's original state house. As the fair's organizers claim, a one-hour trip on foot takes you across all of New England.

STORROWTON VILLAGE
OLD-STYLE CLAM CHOWDER

The members of the Storrowton Village Museum Association prepared a cookbook,
Celebrate the Seasons, *for the bicentennial year 1976. This recipe is adapted from it.*
For a delicious corn chowder, replace the clams with two cups of cooked corn and
a generous pinch of dill weed.

SERVES 4 TO 6

¼ pound salt pork, cut into
    ¼-inch pieces
1 large onion, chopped
1½ cups water
3 medium potatoes, peeled and
    cut into ½-inch dice

4 cups chopped clams
3 cups milk
4 tablespoons (½ stick) butter
    Salt and freshly ground pepper to taste

1. In a heavy skillet, cook the salt pork over moderate heat until browned, turning often. Remove with a slotted spoon and set aside. Over moderate heat cook the onion in the pork fat, stirring frequently, until golden brown.
2. Transfer the onion to a large saucepan. Add the water and the diced potatoes. Bring to the boil, lower the heat and cook for 10 minutes, or until barely tender.
3. Drain the liquid from the potatoes into a large saucepan. Add the clams and their liquid. Simmer for 10 to 20 minutes, until the clams are tender. Add the milk, the potatoes and onion, the butter and the reserved pork crisps. Heat gently. Taste and adjust the seasoning with salt and pepper.

TWO

MIDATLANTIC

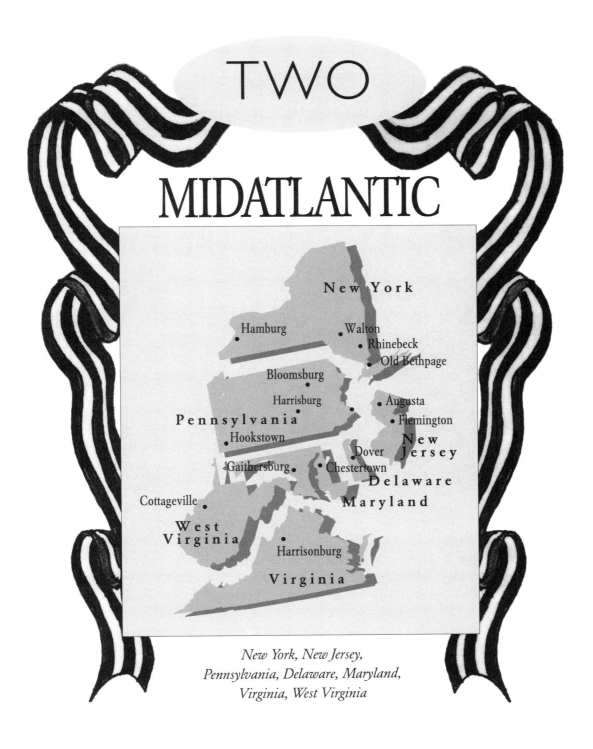

New York
  Hamburg   Walton
             Rhinebeck
             Old Bethpage
    Bloomsburg
    Harrisburg   Augusta
Pennsylvania   Flemington
  Hookstown   New Jersey
             Dover
  Gaithersburg   Chestertown
                      Delaware
Cottageville   Maryland
West
Virginia
             Harrisonburg
             Virginia

New York, New Jersey,
Pennsylvania, Delaware, Maryland,
Virginia, West Virginia

LONG ISLAND FAIR

Old Bethpage, New York

FAIR DATE: 8 days in mid-October

GAZETTEER: *The Long Island Fair is sponsored by the Agricultural Society of Queens, Nassau and Suffolk Counties, and the County of Nassau. The site is Old Bethpage village restoration, Round Swamp Road, Exit 48, Long Island Expressway, Old Bethpage, Long Island.*

Until World War II, Long Island, the one-hundred-and-eighteen-mile-long spit of land to the east of New York City, was largely bucolic—farms, grazing land, market gardens. But at war's end returning veterans and their brides, armed with the G.I. Bill and lured by the new affordable tract housing, turned much of the countryside into a vast suburb. Malls have replaced potato fields, and most of Long Island's young people are oblivious to the recent rural past.

Until, that is, they visit the Old Bethpage village restoration, a mid-nineteenth-century living museum farm village, during the annual eight-day run of the Long Island Fair. The fair's organizers have rigorously stripped away all vestiges of twentieth-century honky-tonk; no Ferris wheels, no fortune games, no commercial displays. "We've returned to our original purpose and spirit," says Dorothy V. Reinhard, the fair's manager. "We're an agricultural fair, with competitions suited to today, and entertainment appropriate to the 1870s. Our volunteers dress in period costumes, so the illusion is perfect." Indeed it is. Blue-clad veterans of the Grand Army of the Republic (actually a group of military re-enactors in Civil War uniforms) reminisce about their "days at the front," and jugglers and magicians prove that they can captivate a late-twentieth-century crowd.

The Long Island Fair has run almost every year since 1842, making it one of the oldest agricultural exhibitions in the United States. The fair has shifted grounds a number of times since its inception, but the sponsors found the ideal, idyllic site on the fairgrounds of the village restoration, a "community" of historic houses, churches, a schoolhouse and farm buildings moved from original sites around Long Island to a two-hundred-year-old farm (only one mile from the ever-jammed Long Island Expressway) that was saved from developers by concerned citizens who saw the need to preserve the area's history.

G.A.R. Civil War re-enactor.

The fair is about as close as contemporary visitors can come to the feel of a nineteenth-century agricultural observance; there is an eighth-of-an-acre horse plowing contest (first prize, ten dollars) stressing "straight rows, depth and evenness of cut." And though such cross-cultural twentieth-century entries as oriental vegetables and ecology boxes are encouraged, so are competitive displays of historic fruit and vegetable varieties. All entertainment and agricultural judging is held under large blue and yellow striped tents. The fairgrounds buildings are re-creations, faithful to old photographic evidence of earlier fair buildings.

The heart of any agricultural fair is the judging: swine, cattle, sheep, dairy goats and draft horses; vegetables; trays of assorted Long Island potatoes, Chinese vegetables (minimum: eight varieties), cauliflower (minimum: 3 varieties), on through beets and mangel-wurzels, carrots, both cylindrical and tapered, and of course tomatoes, tomatoes, tomatoes. Some classes are open to farmers, others to home gardeners. Growers can compete in thirty or more apple classes, displaying varieties from Baldwin to Wolf River. There are more than twelve peach classes, including Sweet Sue and Late Rose. Needlework is a large division, with fierce competition for quilts, wall hangings and embroidered apparel. And the culinary arts! Jellies, jams, preserves, pickles, breads, cakes and all baked goods imaginable. "No more than

TWO BAKED entries per person," as the rules state. There are blue ribbons and glory (and a small monetary prize) to the winners, a resolve to try harder by the also-rans, and the chance for all to learn.

For farmers, the Long Island Fair is serious business. Two families have exhibited the fruits—and vegetables—of their labors at the fair for over fifty years: the Youngs of the westerly, built-up county of Nassau, and the Wickhams of Cutchogue, farther out on Long Island, in heavily farmed Suffolk County.

THE YOUNGS OF NASSAU COUNTY

John H. Youngs is president of the Agricultural Society, and a vegetable farmer. In Nassau County, where most of the population consists of newcomers, six generations of Youngs have lived on Long Island, and on Hegeman's Lane in Old Brookville for one hundred years.

"All that we grow we sell at our farm stand," says John. "My father was a dairyman, and grew vegetables too for the New York market. He took them in by horse and wagon, and then by truck, five days a week, and sold for what he could get. Often he didn't make cost.

"Since 1960 we've grown only vegetables, sold retail at the stand. We grow vine-ripened greenhouse tomatoes, which we start early," says Vivian, John's energetic wife. "John was part of the group that moved the fair from Mineola to Bethpage. He plants things especially for the fair. He's exhibited for over fifty years."

"I've won prizes for pumpkins, mostly," says John, "but we show many other vegetables." John Youngs and his father developed a popular round buff-colored pumpkin, called "Youngs Beauty." After forty years of selection, there is now a large Halloween pumpkin called "Youngs Pride"; they're proud to have preserved the seed. "Our pies are made from the Long Island cheese pumpkin," says Vivian.

The Youngs have turned the operation of the farm stand over to their daughters, who also have a thriving baked goods business, "with homemade flavor." Jo-Hana Youngs Gooth and Paula Youngs Weir, with the help of four employees, bake up to one hundred pies a day in addition to cases of muffins, breads and cookies. They'll test and change until they get what pleases them. "Most baked goods are too sweet for our tastes," says Jo-Hana, "so we invariably reduce the sugar."

A LONG ISLAND CLAMBAKE

*A Fourth of July tradition since the mid-1800s, the Long Island clambake differs from the better-known New England one, where soft-shelled steamer clams, corn, chickens, potatoes, onions and even lobsters are stacked over glowing coals and seaweed, covered with tarpaulins and steamed until cooked. On Long Island, native hard-shelled clams—quahogs (*Mercenaria mercenaria*)—are used. "Not large chowder clams, not cherrystones," says John Youngs, "somewhere in between."*

For this feast, only the clams are cooked outdoors. The rest of the meal—potato salad, "cabbage salad" (coleslaw), biscuits, lettuce and tomatoes, baked ham and corn—is cooked in the kitchen. "It's early in the year for corn," says John, "but we manage to go out in the field and find enough." Vivian Youngs says that some of the shell disintegrates, and the clams become chewy and smoky. "They need a lot of melted butter," she notes.

Here are John's instructions for a successful Long Island clambake:

"You need the iron rim of an old wagon wheel, about three feet in diameter is good. You line the rim with a thick wad of old newspapers. You arrange the clams, lots of them, inside the rim, and it's important to put them valve side down, so they open downward and lose their juices onto the papers.

"Next, you build a kindling fire over the clams; we use old broken-up boxes. The fire burns only about ten to fifteen minutes, then goes out. The clams cook two ways—partially baked from the fire on top, and steamed by the steam that rises from the wet newspapers. You brush the ashes away, put the clams on platters with tongs, and eat them with lots of butter. They're absolutely delicious. I can taste them right now, thinking about it."

COOKING SWEET CORN

Vivian Youngs has strong opinions when it comes to cooking sweet table corn:

"You husk the corn, put it in the boiling water, let the water come back to the boil, and time it for four minutes. Take the corn out, put it on a platter and cover it with a towel—don't leave it in the water to soak; it just gets tough. Corn, even when cooled, is nice to eat. We put the extra aside in a cool place, and my husband eats it as a midnight snack. In fact, he prefers cooled corn.

"If you do refrigerate cooked corn, cut it off the cob and heat it up next day in butter; it's delicious."

YOUNGS FARM PEACH JAM BREAD

The Youngs grow the peaches for the jam on their land. "It's wonderful for breakfast," says Paula, "or warmed for dessert, with fresh peach ice cream."

MAKES 2 LOAVES

1½ cups all-purpose flour	⅓ cup vegetable shortening
¾ teaspoon salt	2 large eggs
½ teaspoon baking soda	¾ cup peach jam
½ teaspoon ground cinnamon	½ teaspoon vanilla extract
½ teaspoon grated nutmeg	½ cup buttermilk
¾ cup sugar	

- Preheat the oven to 350 degrees.
1. Grease and flour two 7 1/2 × 3 1/2 × 2-inch loaf pans. Set aside.
2. In a large mixing bowl stir together the flour, salt, soda, cinnamon and nutmeg; set aside.
3. In the large bowl of a standing electric mixer beat the sugar and shortening on high speed until light and fluffy. Beat in the eggs, one at a time, beating one minute after each addition. Add the jam and vanilla; mix well.
4. Add the flour mixture and buttermilk alternately to the creamed mixture, beating well after each addition.
5. Pour the batter into the prepared loaf pans. Smooth with a spatula. Bake in the center of the oven for 40 to 45 minutes or until a toothpick inserted in the center comes out clean. Place the pans on a rack for 10 minutes. Remove the bread from the pans and let cool on the rack.

THE WICKHAMS OF SUFFOLK COUNTY

Wickham's Fruit Farm occupies two hundred acres of rich land on the shores of Peconic Bay, some ninety miles from New York City, on the North Fork of Long Island. This is one of the oldest pieces of farmed land in America, and almost certainly one of the oldest farmed by one family; Wickhams have been here for almost three centuries.

Map of Long Island, with Wickham Farm.

Members of the family are partners, but in a generations-old tradition one member is in charge. In 1987 Tom Wickham came home from a twenty-year career in Asia as an irrigation engineer to take over management from his seventy-year-old-father, John.

John Wickham is still active in the family endeavor but it was time to cede his place. "I defer to Tom," says John; "that's the way it has to be." John is a trustee of the Agricultural Society of which his friend John Youngs is president, and he's exhibited at the Long Island Fair for over fifty years. "We compete with everybody," says John's wife, Anne. "Not so much for the competition, but to show that these fruits are still grown on eastern Long Island."

Like the Youngs, the Wickhams sell all that they grow at their farm stand. The roadside market is open from May to December.

Twenty crops in all are grown, but pride of the farm is their highest-quality white and yellow peaches, tree ripened, fresh picked, and hand graded daily. The newest crop is apricots, coming in with the earliest peaches in mid-July.

Difficult as the life is, the Wickhams have never thought of *not* farming. "We've stuck to it through thick and thin," says John Wickham. "I fully expect someone from the next generation and the one after that to keep the farm going." Says his son Tom, "I came back to the farm. There was no doubting that I would."

ANNE L. WICKHAM'S BAKED ASPARAGUS

Mrs. Wickham says, "We open the stand in May, when the first asparagus is ready. Baking it concentrates the flavor. The local asparagus is so tender, we never think of peeling it."

SERVES 4

1 pound fresh asparagus, washed, any tough ends removed Salt and freshly ground pepper to taste	2 to 3 teaspoons butter, softened 3 tablespoons freshly grated Parmesan cheese (optional)

- Preheat the oven to 400 degrees.
1. Place the asparagus in a baking dish that just fits it. Add salt and pepper and dot with butter. Sprinkle with Parmesan, if desired. Cover the dish tightly with foil and bake until tender, about 20 minutes.
2. If you have used Parmesan, remove the foil and brown the dish briefly under the broiler.

WICKHAM FAMILY PICKLED CHERRIES

"No need to process these cherries in a water bath," says Mrs. Wickham. "They'll keep all winter on the shelf. We serve them as an accompaniment to poultry or meat."

Sour cherries (any amount)   Sugar
Cider vinegar to cover

1. Wash, stem and pit the cherries. Place them in a bowl and cover with cider vinegar. Let stand overnight.
2. Pour off and discard the vinegar. Measure the cherries and add an equal amount of sugar to them. Let stand until the sugar dissolves, stirring occasionally. Put the cherries in very clean jars and cover. Store in a cool dark place.

STRAWBERRY OR BLUEBERRY PIE

"You might want to omit the cream cheese if you use blueberries," says Anne Wickham.

MAKES ONE 9-INCH PIE

1 (3-ounce) package of cream cheese, softened   1 cup sugar
1 tablespoon milk or cream   3 tablespoons cornstarch
1 fully baked 9-inch pie shell   1 cup whipping cream
4 cups hulled ripe strawberries or fresh blueberries

1. Mash the cream cheese with the milk until spreadable. Spread it over the pie shell. Arrange half the strawberries, pointed ends up, or blueberries, in the pie shell.
2. Place the remaining 2 cups of strawberries or blueberries in a saucepan. Add the sugar and the cornstarch, pressed through a sieve. Cook, mashing the fruit with a wooden spoon, until the mixture comes to the boil. It will thicken almost at once. Pour while still hot over the fruit in the pie shell. Cool and chill in the refrigerator.
3. Whip the cream and pipe or spread it over the pie.

DUTCHESS COUNTY FAIR

Rhinebeck, New York

FAIR DATE: THIRD WEEK IN AUGUST, 6 DAYS

GAZETTEER: *The fairgrounds are on Route 9, Rhinebeck, a charming town located halfway between New York and Albany. Rhinebeck was first settled in 1686. Dutch patroons and English grandees—Beekmans, Livingstons, Astors and Schuylers—established great estates in the lovely Hudson valley, bordering the majestic Hudson River. Many estates are now open to the public: both Franklin D. Roosevelt's home and library and the Vanderbilt estate are nearby, on Route 9, Hyde Park. So is the Culinary Institute of America ("the other CIA"), training ground of many of America's finest chefs. The CIA runs several dining rooms. Exceptionally good antiquing is found throughout Dutchess County.*

"Ours is a very large fair," says Pat Pflum, a fair employee, "but we are proud to have a parklike rather than a carnival atmosphere. Even the trailers of exhibitors and visitors have hanging baskets of flowers. We have a huge exhibit of livestock, and two thousand plus culinary and home craft exhibits."

The Home Beer Making Competition draws a good number of suds enthusiasts. So does the Home Beer Label Competition. Categories include Ale (light, dark, pale, brown and Scottish bitter are some of the designations); Lagers (light, amber, dark and bock) and Any Other Beer (porter, stout, steam, barley, ginger, wheat, fruit and

mead). Soda labels (sarsaparilla, root beer and others) are judged, too. At only fifty cents per entry, anyone can vie for brew-related artistic achievement.

"When I was a boy, before we went to the fair we'd get the chores done," says Fred Briggs, who grew up near Rhinebeck. "Mother put a picnic in a basket, and we went over early in the wagon. Tied the horse up in the orchard, and we roamed around. My father went to the cattle show—we had a small herd—and Mother took her bread or handiwork to enter it in the competition. For me, the big deal was the Ferris wheel; I only saw it once a year." Now Mr. Briggs, an antique postcard buff, is in charge of the entries in the antiques section, a unique feature of the Dutchess County Fair. "We take bona fide family heirlooms for display, showing how people in Dutchess lived a long time ago," he says. "A girl might make a patchwork quilt and show it alongside of one her great-grandmother made one hundred years ago. We give prizes for the best in some twenty categories, including cut glass, coins and sterling silver. The heirlooms go home with their owners at the end of the week."

CATHY POLUZZI AND NORA MERRIAM

Cathy and Nora collaborated on these recipes, designed to highlight Dutchess County's exemplary apple and beef. Cathy, who has been with the fair since 1978 and with newcomer Nora is co-chair of the Culinary Unit, is a fine baker. Cathy's daughter Angela is also a talented baker; Angela won the baking prize for children at the ripe age of five.

RUSSIAN BLACK BREAD WITH HERB BUTTER

"Chocolate and instant coffee may not, strictly speaking, be Russian, but they contribute to the great taste of the bread," says Cathy. "Sensational with herb butter!" Nora adds.

MAKES TWO ROUND LOAVES

4 cups rye flour	Water
2 cups whole bran cereal	4 tablespoons (½ stick) butter
2 packages active dry yeast	¼ cup white vinegar
2 teaspoons instant coffee	¼ cup dark molasses
2 teaspoons granulated sugar	1 ounce (1 square) unsweetened dark
1 teaspoon brown sugar	chocolate
1 teaspoon salt	2½ to 3 cups unbleached all-purpose flour
½ teaspoon fennel seeds, crushed	1 teaspoon cornstarch
2 tablespoons caraway seeds, crushed	

1. Lightly grease a large bowl and set aside. In a separate large bowl, combine the rye flour, bran, yeast, coffee, granulated sugar, brown sugar, salt, fennel and caraway seeds and mix well.

2. In a 2-quart saucepan, combine 2 1/2 cups of water, the butter, vinegar, molasses and chocolate. Stir over moderate heat until the chocolate is almost melted but the mixture is still lukewarm. Add to the rye flour mixture and begin beating with a wooden spoon. Add the all-purpose white flour gradually, 1/2 cup at a time, making a soft dough. Beat about 3 minutes.

3. Turn the dough onto a lightly floured surface and cover with a large inverted bowl. Let the dough rest 10 to 15 minutes. Knead until smooth and elastic, about 10 minutes, adding additional white flour as needed. Place in the greased bowl, turning to coat the entire surface. Cover the bowl with plastic wrap and a hot damp towel: leave in a warm place until doubled in bulk, about 1 1/2 to 2 hours.

4. Punch the dough down and turn out onto a lightly floured surface. Shape into 2 balls and place in 2 greased 8-inch cake pans. Cover lightly with greased plastic wrap and let rise in a warm place until doubled, about 1 hour.

• Preheat the oven to 350 degrees.

5. Bake the loaves for 40 minutes. Meantime, combine the cornstarch with 1/2 cup of water in a saucepan and bring to the boil. Boil over high heat for 1 minute. Brush the mixture over the loaves and return them to the oven for 5 minutes, or until the tops are glazed and the loaves sound hollow when tapped. Remove from pans and cool on racks.

HERB BUTTER

Nora and Cathy agree: "Slather this butter on the Russian bread—in fact, on any bread!"

MAKES 1/2 CUP

8 tablespoons (1 stick) butter,
    at room temperature
⅛ teaspoon crumbled thyme
⅛ teaspoon garlic powder
⅛ teaspoon crumbled oregano

2 tablespoons finely chopped
    fresh parsley
1 teaspoon finely chopped shallots
2 teaspoons grated lemon peel

- Mix all ingredients, mashing with a wooden spoon, until thoroughly combined. Allow to stand in a cool place for several hours, so that flavors blend and mellow.

APPLE AND SAUSAGE PASTRY ROLL

"We love this on a brisk fall day," Cathy says.

SERVES 6

2 cups all-purpose flour
½ teaspoon salt
1 tablespoon baking powder
⅓ cup vegetable shortening
⅔ cup milk
1½ pounds hot or sweet sausage,

without casings, crumbled, cooked
    and drained
2 cups chopped peeled apples
    (2 medium apples)
2 teaspoons paprika
    Dash of ground red pepper

1. Sift the flour, salt and baking powder into a mixing bowl. Cut in the shortening until the mixture resembles coarse meal. Add the milk and stir just until combined; the dough will be soft.
2. Turn the dough out onto a floured work surface. Knead gently a few times, then wrap in plastic and chill for 30 minutes.
- Preheat the oven to 350 degrees.
3. Place the cooked sausage in a mixing bowl. Mix in the chopped apples, paprika and red pepper.

4. Roll the pastry out on a floured surface to a rectangle about 15 × 9 inches. Cover it evenly to within 1 inch of the edges with the sausage mixture. Roll up, jelly roll fashion, starting at the short end. Pinch the edges together and moisten lightly with water to seal in the juices. Place the roll on a baking sheet. Bake 20 to 25 minutes, until nicely browned. Let cool somewhat before slicing.

SPIT-ROASTED BEEF RIB ROAST

You will need a grill with a rotisserie attachment for this recipe.
Nora says, "Experiment with different woods or chips, such as mesquite or hickory."

SERVES 6 TO 8

1 (3½-pound) boneless beef rib roast	2 teaspoons freshly ground black pepper, or more to taste
1 tablespoon seasoned salt, such as Lowry's	½ teaspoon garlic powder

1. The meat must be at room temperature. Rub it well all over with the combined seasonings. Insert the spit prongs firmly at both ends of the roast, and test for balance by rotating the spit on the palms of both hands.
2. Build a fire at the rear of the grill. If using briquettes, they should reach the gray-ash stage. Place the drip pan in front of the fire and under the roast.
3. Attach the spit and start the motor. The slow, steady turning allows the meat to baste itself. Roast for 45 to 60 minutes, or until the roast is done to your liking.
4. Remove from the spit and let the roast rest 20 minutes before slicing.

DELAWARE COUNTY FAIR

Walton, New York

FAIR DATE: **MONDAY THROUGH SATURDAY, THIRD WEEK IN AUGUST**

GAZETTEER: *Walton is in the western Catskill Mountains, 3 hours from New York City. Coming from New York, take Route 17 to Roscoe, then take 206 to Walton. Though dairy farms form the economic base of this countryside of rolling hills and river valleys, much land has been sold for the development of vacation houses. Cooperstown, with the Baseball Hall of Fame, is fifty miles from Walton.*

As state funds became available to agricultural societies sponsoring fairs, scattered town and village gatherings began to coalesce into larger county fairs. In 1887 the Delaware Valley Agricultural Society was formed, to sponsor a fair in Walton.

"My husband's father, Alex Tweedy, was a member of the first board of the Walton Agricultural Fair," says Catherine Russell. "We remain a fair family. My son Renwick, who teaches agriculture, is adviser to the Future Farmers of America at the local high school. At the fair he exhibits hay, corn and other FFA projects. Renny's wife, Candace, is superintendent of the building with the flowers, craft work, table settings, and baked and canned goods.

"We run a dairy farm. Years ago my husband's father went to Wisconsin to buy healthy cows after his herd tested positive for TB. He came back on the train, riding alongside twenty cows, and got off at a station two miles from here. He unloaded the cows and drove them home to the barn.

"Over eighty-five years ago, when my husband, Robert, was born, his mother got the idea of having city people come to stay on the farm. The farmhouse had ten extra bedrooms. She got acquainted with a woman from New York City through our church, and there's not a summer since that we have not had people here. We've formed friendships, and we've learned from each other. A man who'd been here as a boy brought his children. His little girl hurt her feet on the steps; she'd only ridden elevators, and wasn't used to stairs. Those folks are still coming. We had two little boys arrive wearing bicycle helmets, to protect them from the wild animals! They learned to love the animals, to appreciate the carrots out of the ground, and seeing the cows milked. Most children are surprised that the milk is warm; they've only had milk ice cold.

Russell's Wayside Rest.

"City people, used to light, can't believe how dark night is in the country. One fellow woke up and thought he'd lost his sight! Now we have a security light at the corner. The quietness of the country impresses people greatly, a pleasure we take for granted. But one lady, used to noise and telephones, was not happy at all.

"These days hunters come here, too. We have more than four hundred acres they can roam over. They're no trouble to me, I just give them their breakfast at five a.m.—coffee, juice, hot cereal, toast, bacon and eggs or ham—and they get their other meals in town.

"Our place is called Russell's Wayside Rest. It was good for our children growing up to make friends with city children. But we're getting up in years now, and Wayside Rest is coming to an end."

CATHERINE RUSSELL'S MAPLE MOUSSE

Many sugar maples grow in New York State; Russell's Wayside Rest has its own stand of maple trees. This elegant dessert could be topped with fresh raspberries.

SERVES 8

3 cups milk	1 cup maple syrup
4 egg yolks	4 egg whites
2 cups sugar	2 cups heavy cream
2 tablespoons flour	

1. In a large saucepan, scald the milk by heating it just until little bubbles form at the edge of the pan. Do not let it boil. Remove from the heat.
2. Beat the egg yolks well. Mix 1 cup of the sugar with the flour. Add the maple syrup, the sugar-flour mixture, and the beaten yolks to the saucepan of milk. Cook over low heat, stirring constantly, until the mixture thickens enough to coat a wooden spoon, about 10 minutes. Remove immediately from the heat and let cool. Cooling takes 30 minutes or longer.
3. In a bowl, beat the egg whites until soft peaks form. In another bowl, whip the cream, slowly adding the remaining cup of sugar, until soft peaks form. Fold the egg whites into the whipped cream.
4. When the custard has cooled, fold the egg white–cream mixture into it. Spoon the mixture into 2 refrigerator trays and place in the freezer. Stir the mixture twice as it freezes, to break the crystals. Freeze until firm.

GRANDMA MOSES, INSPIRED PAINTER AND EXCELLENT COOK

Anna Mary Robertson Moses (1860–1961), a farm woman who is probably the most celebrated naïf painter of all time, was an enthusiastic fair exhibitor. She won many blue ribbons for her jams and jellies, but, ironically, received no notice at all for the paintings she exhibited at fairs.

Grandma Moses grew up in Eagle Bridge, New York. At age eleven she went to work as a hired girl; she eventually married the hired hand, Thomas Salmon Moses, and the young couple went to Virginia to farm.

"She felt strongly that husband and wife should both pull their own weight," says Jane Kallir, co-director of the Galerie St. Etienne, which represents the Moses estate. "In Virginia, she made butter and sold it, and also made and sold potato chips; she would take them to the general store, and receive payment either in cash or in trade."

The Moses family moved back to Eagle Bridge. "As she aged, she got rheumatism and couldn't use her hands as she had," says Carl Moses, a retired farmer and Grandma's grandson. "She first made yarn stitchery pictures, then went to painting when she was around 72, I think. Until her paintings were discovered, she gave most of them away."

"Around Easter of 1938," says Jane Kallir, "a traveling state engineer named Louis Calder saw her paintings displayed in a drugstore in Hoosick Falls, New York, near her home. He bought them all, and encouraged her to take art seriously. He tried to get New York galleries interested in her, but they balked at her age, almost 80. But my grandfather, Otto Kallir, immediately recognized her talent and gave her an exhibition in 1940, 'What a Farm Wife Painted.' She declined a trip to the opening, saying she'd already seen the paintings, but spoke to a large crowd when some of the pictures were shown at Gimbel's department store in New York City at Thanksgiving. She talked very naturally about her jellies and jams, and won the hearts of all. She didn't realize she was speaking into a microphone, never having seen one. That was the beginning of the legend. For the last twenty-one years of her life she enjoyed success and celebrity that no artist had heretofore received."

"Moving Day on the Farm."

Alas, we do not have records of Grandma's preserves or her potato chips, but we do have some fragmentary handwritten recipes. "Mrs. Whiteside, for whom she worked as a hired girl, gave her a journal in the 1870s, with a few recipes written in it," says Ruth Levin, registrar of the Bennington Museum in Bennington, Vermont. "We own that journal. Throughout her life Grandma added her own recipes, and filled the pages with recipes torn from magazines and newspapers. We also exhibit the largest number of her paintings of any museum, and in the little schoolhouse she and many of her progeny attended, which is part of our museum, we play a 1955 taped interview with Edward R. Murrow made when Grandma was an active and alert ninety-five. We also have her Bible, glasses, and a wicker chair."

GRANDMA MOSES' MACAROONS

Here is Grandma's recipe, just as she wrote it in her journal:

White of 9 eggs, beat to a froth
1 lb blanched almonds, pounded
½ lb sugar

Nutmeg and mace to your taste
To be baked 1 hour on paper

If you are game to try making Grandma's Macaroons, do the following:

• Heat the oven to 325 degrees. Line baking sheets with parchment paper.
1. Combine the blanched almonds and sugar, and blend with 1/2 teaspoon each of freshly grated nutmeg and pounded mace. Fold in the egg whites. Place the mixture in a pastry bag and squeeze out small drops about 1 inch across, placed 2 inches apart on the paper. Bake for 30 minutes. When cool, dampen the underside of the paper with a moist cloth and remove the cookies.

Another recipe:

PLUM PUDDING

½ pound of raisins, the same of suet, currants & flour. ½ pint of milk & 4 eggs. To be boiled four hours.

ERIE COUNTY FAIR & EXPO

Hamburg, New York

FAIR DATE: 11 DAYS, BEGINNING SECOND THURSDAY IN AUGUST

Hamburg is off exits 56 and 57 of the New York State Throughway, 50 miles from Niagara Falls and 20 miles south of Buffalo. Near the Finger Lakes Recreational Area. Home of the Buffalo Raceway.

"We like to say we're the largest ten-day county fair in the United States," says Lloyd Lamb, the fair's general manager. "More than half the people living in western New York visit us."

The Agricultural Society held its first fair in 1820, and today the goals are the same: to promote and encourage agriculture in western New York, still a major farming region. (The fair's huge Agri-Center is open for livestock and horse events year round.) Families picnic at four parks on the grounds, and nine thousand fans can watch the harness races from the grandstand. "We have a host of free shows," says Lamb. "People can enjoy themselves all day long without spending much."

DOLLY'S HOT CHILI SAUCE

"I took a blue or red ribbon with this sauce every year for ten years," Dolly Kress says. "Most chili sauces are dark, because the sugar cooks a long time and caramelizes. I put the sugar in late, and the sauce stays bright red."

MAKES 6 PINTS

20 large ripe tomatoes, peeled and chopped (about 12 cups)	5½ cups distilled white vinegar
6 large onions, ground	3 tablespoons pickling salt
12 hot peppers, seeds removed, ground (about 2 cups) (see Note)	4 cups sugar

1. In a large heavy nonaluminum kettle, cook the vegetables with the vinegar and salt over moderate heat, uncovered, until reduced to about half, about 2 hours.
2. Add the sugar and cook until thick, about 1 hour, stirring frequently. The sauce should be bright red. Pour into hot sterilized jars, seal, and process in a boiling water bath for 10 minutes.

NOTE: If you want a less fiery sauce, use only a few hot peppers and substitute ground sweet red bell peppers for the rest of the 2 cups.

1821 1971

★ THE GREAT ★ ERIE COUNTY FAIR & EXPO

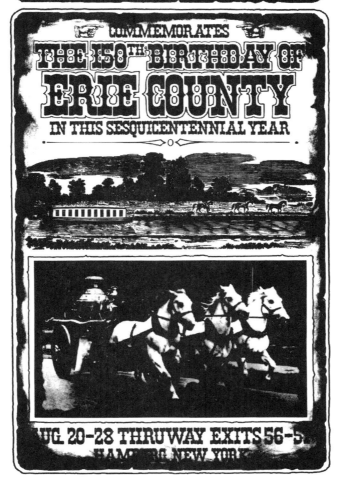

COMMEMORATES
THE 150TH BIRTHDAY OF
ERIE COUNTY
IN THIS SESQUICENTENNIAL YEAR

AUG. 20-28 THRUWAY EXITS 56-5
HAMBURG, NEW YORK

BUTTER HORNS

"My family has made these for years, especially at Christmas and Easter. I am Polish, and my mother recalls her mother made something similar. They aren't quick and easy, but the end result is worth the effort," Dolly says.

MAKES 72 BUTTER HORNS

THE DOUGH:

1 package active dry yeast	3 sticks unsalted butter, cut into bits
¼ cup warm water	2 egg yolks, beaten
4 cups all-purpose flour	½ cup sour cream
½ teaspoon salt	1 teaspoon vanilla extract

THE FILLING:

3 egg whites	1 teaspoon vanilla extract
¾ cup granulated sugar	Confectioners' sugar, for dusting
1 cup nuts, finely chopped	

1. Dissolve the yeast in the warm water; in 5 to 10 minutes the yeast should foam. Whisk the flour with the salt in a large mixing bowl. Cut in the butter with a pastry blender or 2 knives until the mixture resembles coarse crumbs. Blend in the dissolved yeast, the egg yolks, sour cream and vanilla. Wrap the dough in plastic and refrigerate for at least 1 hour.
2. Make the filling. Beat the egg whites with the granulated sugar until soft peaks form. Fold in the nuts and vanilla.
- Preheat the oven to 375 degrees.
3. Divide the dough into 12 equal portions. Roll each portion into a thin 8-inch circle. Cut each circle into 6 pie-shaped parts.
4. Place a tablespoon of filling on each wedge and spread slightly. Starting at the large end, roll up and place on lightly greased baking sheets: with your fingers shape each portion into a crescent. Bake about 10 minutes, or until lightly browned. Cool on a rack and dust with confectioners' sugar while still warm.

THE BURGER BOYS

Where did that American culinary icon, the hamburger, first appear? The Erie County Agricultural Fair has a credible claim. The year was 1885, and the Menches brothers, Charles and Frank, arrived to sell their specialty, pork sausage sandwiches. New regulations were in effect: no wood or coal fuel was to be used by the vendors of hot food in the center of the fairgrounds. The pancake man, the hot potato man, the stew and chowder men and the sellers of fried mush sandwiches were relegated to the far edges of the site. The reason? The ladies would no longer tolerate the clouds of smoke that smudged their white dresses.

However, the Menches had a newfangled, clean-burning gasoline stove, and thus a prized spot near the new grandstand. But on opening day, no pork was available. Rescue came from Andrew Klein, a resourceful butcher wielding two cleavers, who reduced five pounds of beef to a fine chop. The brothers found that a little brown sugar sprinkled on the griddle gave the patties a full flavor. The customers thought so, too. The sandwiches disappeared quicker than Frank could say "hamburger," which is what he named the creation, in honor of the Erie County Fair, also known as the Hamburg Fair, after the town where it is held.

SUSSEX COUNTY FARM & HORSE SHOW

Augusta, New Jersey

FAIR DATE: GENERALLY, THE FIRST WEEK OF AUGUST, LASTING 10 DAYS

GAZETTEER: *Sussex is still largely rural. The terrain is hilly, with many lakes. The area has several important amusement parks and a game preserve. Coming from New York, take Route 80 to Route 15. At 15, bypass around Sparta to 206 North. You will see signs.*

"Sussex is one of the larger fairs in the Northeast," says Warren Welsh, who has been with the fair since 1950 and is its secretary. "We've had events like cow-chip throwing contests; milking contests, like the one some years ago, which was politicians against Playboy Club bunnies; and ice cream eating contests," says Welsh. "A really major horse show is part of our attraction. We also have spectacular rigs like the Coors Belgian Draft Horse Team and the Budweiser Clydesdales."

In the ring.

"Caring people in the community make the fair special," says Jane Brodhecker, who coordinates the food booths. With her husband Tom, a retired airline pilot and vice president of the fair, she runs the family farm, raising sheep, beef and grain. "We're all volunteers here, and our general interests and goals are the same. We're all friends throughout the year," Jane says.

At Sussex, New Jersey justifies its nickname, "The Garden State." "It's the opposite of what people think New Jersey is," remarked one fairgoer. "Wall-to-wall traffic and hurrying people."

JANE BRODHECKER'S LAMB QUICHE

"We raise lambs for the spring market," says Jane, "but they go so fast we don't get much ourselves. The quiche is one of our favorites. We like it for buffets, and we served it at our daughter Patti's wedding brunch."

SERVES 6

1 pound ground lamb	1 teaspoon salt
⅓ cup chopped onions	¼ teaspoon freshly ground pepper
4 eggs, slightly beaten	An unbaked pie shell
1½ cups grated cheddar cheese	3 tablespoons minced parsley
½ cup milk	Paprika, for sprinkling

- Preheat the oven to 425 degrees.
1. In a skillet, cook the lamb over moderate heat, stirring, until it loses its pink color. Add the onion and cook until it is soft and translucent and the meat is lightly browned. Pour off all fat.
2. In a bowl, combine the eggs, 1 cup of the cheese, the milk, salt and pepper. Add the meat-onion mixture and pour into the pie shell. Top with the remaining cheese and the parsley. Sprinkle with paprika. Bake in the center of the oven 40 to 45 minutes, until the cheese is bubbling and the pie has begun to set like a custard. Serve warm.

HOT SPICY CANDY

"The recipe comes from my Indiana grandmother," says Jane. "We made this candy with our children, and now they make it with theirs. Before Christmas everybody comes to the farm and we make batches together in several flavors. We use red coloring for cinnamon candy, black for anise, and purple for spearmint. Then we mix them and put them in airtight containers for gifts. The procedure sounds complicated, but making this candy is really simple."

Flavored oils—cinnamon, anise, wintergreen and others—can be found at many pharmacies and sometimes at specialty food shops. Intense fruit flavorings are good, too. Paste coloring agents are sold at bakers' supplies stores and craft shops.

MAKES ABOUT 4 CUPS

3¾ cups granulated sugar
1½ cups light corn syrup
1 cup water
Paste or liquid pure food coloring

1 dram (scant teaspoon) oil of cinnamon, clove or peppermint, or fruit flavoring
2 pounds confectioners' sugar (see Note)

1. Place the sugar, syrup and water in a large saucepan. Bring slowly to the boil, with a candy thermometer attached to the side of the pan. Heat the mixture to exactly 290 degrees. (This will take 15 to 20 minutes.) Higher, the candy sets too hard; short of 290, it won't set up. Remove from the heat and add the desired color. Paste colors are intense; a dab on a toothpick should give a deep color. Be careful with liquid colors—too much may dilute the candy mixture and impede setting.
2. Stir in the oil flavoring. Stand back! The mixture steams up and releases strong fumes.
3. Have ready 3 jelly roll pans on which you have placed a 3/4-inch-thick fluffy layer of confectioners' sugar. With a finger, trace a large spiral trough in each pan.
4. Carefully pour the liquid into the troughs—the little walls of confectioners' sugar keep it from spreading. When the candy has hardened and is cool enough to touch—a matter of minutes—take scissors and snip it into short pieces, or snap off pieces with your fingers. Roll them in the sugar. Sealed in containers, the candy lasts for months.

NOTE: You can reuse the confectioners' sugar. Between candy-making bouts, store the sugar in a self-sealing plastic bag.

FLEMINGTON AGRICULTURAL FAIR

Flemington, New Jersey

FAIR DATE: FULL WEEK BEFORE LABOR DAY

GAZETTEER: *Located on Route 31, the fairgrounds and speedway lie 30 minutes north of Trenton, near Flemington and Liberty Village, a large collection of outlet stores. The surrounding countryside is still largely rural.*

Flemington's farsighted organizers know well that the heart of any fair is the quality of its agricultural displays, and the celebration of the fruits of the earth and the labor of the farmer. As family farms increasingly disappear and urban dwellers know less and less about how food is produced, dedicated people like Kathy Kovacs, superintendent of agricultural displays, work to keep the agricultural spirit thriving. Kathy, a dairy farmer's daughter, straddles the older and the newer worlds: "The dairy went years ago," she says. "It's just too much work for too little money. The first time my five-year-old saw a cow being milked was at the Philadelphia Zoo! My father was appalled."

Kathy grew up with the Flemington Fair. "I still see the fair as agricultural. The size, the thrilling rides are a real part, but to me that's largely irrelevant. I can get that at an amusement park. It would be easy to say the fair gets no money from agriculture and lots from concessions, but if we ran the fair for purely financial gain, we'd lose the history. I took over the fruits, vegetables and grain display, and to me it's a personal crusade. I want to see that people can still experience that old-time country camaraderie, the friendly competition: 'my tomatoes can beat your tomatoes' sort of thing. I'm determined to hold onto that. Flemington Fair visitors are getting a piece of Americana that we can continue in the same spirit, by encouraging people who have gardens to grow all kinds of things. It's important for them to know that, young or old, the competition is for them."

MARIE E. JOHNSON'S
CRANBERRY SOUFFLE SALAD

Kathy Kovacs says her friend Marie's holiday side dish is wonderful. "It's colorful and not too sweet," says Marie. "I serve it on lettuce, or some sort of greenery. It can easily be doubled. It's a fine accompaniment to turkey or ham."

SERVES 6

1 envelope unflavored gelatin
2 teaspoons sugar
¼ teaspoon salt
1 cup very hot water
½ cup mayonnaise
2 tablespoons freshly squeezed lemon
    or orange juice

1 teaspoon grated lemon or orange rind
1 (16-ounce) can whole cranberry sauce
1 apple, peeled and sliced
¼ cup chopped walnuts

1. Mix the gelatin, sugar and salt in a large mixing bowl. Add the hot water and stir until the ingredients dissolve. Stir in the mayonnaise, lemon juice and rind. Blend together with a rotary or electric beater. Pour into a metal refrigerator tray and quick-chill in the freezer for about 15 minutes, or until cool and slightly syrupy.
2. Pour into a bowl and beat until fluffy. Fold in the cranberry sauce, diced apple and nuts. Pour into an oiled 6-cup mold or dish and chill 6 hours or overnight until firm.

THE UNDER-THE-FENCE- FRATERNITY

Over twenty-five years ago, the Flemington Fair office received a letter containing fifty cents. This was conscience money from an eighty-five-year-old man, who confessed to crawling under the fence as a boy to avoid paying. Local columnist Dereck Williamson confessed that he was a fence wriggler, too. One kid would dig a hole, the word would spread, and till the guards plugged up the gap many a boy had slid under. "In those days," says Williamson, "if you saw a happy-looking kid sauntering along the midway, dirty like a pigpen, it was a good bet he had not paid the price of admission."

ALICE EVERITT'S QUICK STICKY BUNS

"My mother is famous for her wonderful sticky buns," says Kathy Kovacs. "They're always the first things sold at her church's bake sales, and sometimes they're spoken for before they get there!"

MAKES 18 STICKY BUNS

THE BUNS:

3¼ cups all-purpose flour	4 tablespoons (½ stick) butter
2 packages active dry yeast	1 teaspoon salt
¾ cup milk	¼ cup granulated sugar
½ cup water	1 large egg

THE TOPPING:

12 tablespoons (1½ sticks) butter, at room temperature	1 teaspoon ground cinnamon
1 cup light brown sugar, firmly packed	1 tablespoon light corn syrup
¾ cup chopped walnuts or pecans	1 tablespoon water

1. In the large bowl of a heavy-duty standing electric mixer, put 1 1/2 cups of the flour. Add the yeast and blend with a whisk. In a saucepan, gently heat the milk, water, butter, salt and sugar until little bubbles appear around the rim; do not let boil. Pour over the flour-yeast mixture. Add the egg and beat for 30 seconds at low speed, scraping the bowl constantly. Then beat 3 minutes at medium speed.

2. Gradually add the remaining 1 3/4 cups flour, beating at low speed until the flour is absorbed. Scrape the sides of the bowl. Cover the bowl with oiled plastic wrap and let rise in a warm place until doubled, 30 to 40 minutes or longer.

3. While the dough is rising, prepare the topping. Mix all the ingredients together. Have ready a greased 9 × 13 × 2-inch pan. Spread the topping over the bottom.

4. Punch down the dough; it will be sticky. Drop it by the large spoonful onto the topping; don't worry if the dough balls do not touch; they will as they rise. Cover with oiled plastic wrap and let rise again in a warm draft-free place until doubled, about 30 minutes or longer.

• Preheat the oven to 375 degrees.

5. Bake the buns for 20 to 25 minutes. When golden brown, remove and cool on a rack for 3 minutes. Invert the buns onto a tray or baking sheet. While still warm, cut into 2 × 3-inch portions. Serve the buns, topping side up, warm or at room temperature. They may be reheated.

BLOOMSBURG FAIR

Bloomsburg, Pennsylvania

FAIR DATE: THIRD SATURDAY AFTER LABOR DAY FOR 8 DAYS

GASETTEER: *Bloomsburg calls itself "Gateway to the Poconos," the nearby resort mountains. It is midway between Harrisburg, the state capital, and Wilkes-Barre. Take Route 11 North from Harrisburg to Route 80.*

"Our fair began in 1855," says Fred Trump, the president, "and we've been through some interesting weather since then. But Hurricane Eloise, in 1972, was the challenge. The fairgrounds were seven feet under water; we're right on the banks of the Susquehanna. I'm proud to say that the thousand-some animals were all safe, except for two pigs that jumped out of a boat and drowned.

"Our fair is the largest in Pennsylvania; it's larger than many state fairs. We get about three-quarters of a million people. It's known as an eating fair—people say, 'We love Bloomsburg for the food.' We have a large Amish population that moved up here in Columbia Country, and we have shoofly pie and pickled cabbage in abundance. We also have manicotti, Spanish food, Polish pierogis, and even Cajun food!"

"And we have the oldest dog show in America," says John Brokenshire, who is in charge of it. "We have breeds that most people around here haven't seen, like Irish wolfhounds and bulldogs; fifty or sixty breeds altogether. It isn't a point show, but we do give ribbons. The dogs are here for a week, in kennels. Of course, if one gets too homesick, we have to send it home."

"Part of my job as an extension home economist was giving demonstrations each year at the fair and giving information to the various Granges on what to exhibit," says Anna Mae Lehr. "I learned that folks in this area are interested in having a good time, good food, and a good family life. People here raise beef, hogs, ducks and chickens, and we have a lot of hunting of wild turkey, rabbit and squirrels. Before freezers came along, they had to learn how to can these meats, and we still have hundreds of jars of sausage and spareribs and other meat, especially at Grange exhibits.

"My mother sometimes put up six hundred jars of canned meat. I have fond memories of making *panhaus*, what you'd call scrapple. At hog butchering time we even filled the pig's stomach with mashed potatoes, celery and onions, and baked it. We never wasted a thing. Neighbors would help, and it was a festive occasion.

We'd can some meat, give some away. When my parents passed away, we donated our huge iron kettle and butchering equipment to the farm museum at the fair.

"We also had a party when we made apple butter—we called it a schnitzen party. The night before, we prepared the apples. Some would peel, others slice. The kitchen table was festooned with peelings. Next day the apple butter was cooked outside in that same black kettle, with a very long stirrer so that the one stirring didn't have to stand too near the intense heat from the fire. Sometimes we'd cut sassafras bark from the trees and put it in for flavor. That was back in the 1930s. We made all our own entertainment. People don't do enough of that today. We get so involved that we forget about friendships.

"How we looked forward to fair week! We'd make new dresses and take our fried chicken dinner. I looked forward to the grandstand show. They had a chorus line and dancers from New York. I once stayed with two old ladies who also put up a few of the dancers, and when they found out what the girls did, they had a fit! Now we have Garth Brooks."

VENISON SCRAPPLE

"This is the way they made scrapple in the old days," says Anna Mae Lehr. "It's a very old recipe. You can use pork, too. You will need a big iron kettle, like the one we donated to the Farm Museum at the Bloomsburg Fair. Good luck!"

Save bones from butchering deer or pig. Boil bones until meat falls off, approximately 4 hours. Remove all bones and pick meat off. Strain broth left in kettle and clean out all remaining bones and meat from kettle. Then put broth back in and bring to a boil. Take meat picked off of bones and put in meat grinder. When broth starts boiling, add salt and pepper to taste. Then add the meat. Let kettle come to the boil again, stirring occasionally to keep from burning. When kettle comes to a boil add a 5-pound bag of wheat flour. Keep stirring. When pot comes to a boil again, add 10 pounds of cornmeal. Keep stirring. When pot starts boiling again, add buckwheat flour until thick. Keep stirring the whole time. When thick put in scrapple pan and let cool overnight. When using venison, should put some pork fat in to boil and then grind up meat.

ANNA MAE'S MICROWAVE SPINACH BALLS

In her lifetime, Anna Mae Lehr has made the technological leap from a wood-burning stove to the microwave oven. "As extension home economist, I gave microwave demonstrations at the fair, in a kitchen above the fairgrounds office," Anna Mae says "I like the microwave for some things, like these spinach balls, which I make ahead and serve often as appetizers."

MAKES 24

1 (10-ounce) package of frozen chopped spinach	¼ cup dry bread crumbs
	1 tablespoon grated onion
¾ cup shredded Swiss cheese	1 egg, beaten
2 tablespoons grated Parmesan cheese	½ teaspoon salt

1. Place the package of spinach in the microwave oven. Cook on High 4 to 5 minutes, or until defrosted. Drain, pressing down hard to extract as much liquid as possible.
2. Mix the spinach well with the Swiss and Parmesan cheeses, bread crumbs, onion,

egg and salt. Shape into 1-inch balls, using $1^1/2$ teaspoons of mixture for each. Place on a wax paper-lined baking sheet and cover with plastic wrap. Freeze overnight. (To keep longer, pack a dozen balls in a freezer container and label. Freeze no longer than 2 weeks.)

3. To serve, place the balls on a microwavable paper towel-lined baking sheet. Microwave on High for 2 minutes. Reduce the power to half (Medium). Microwave $4^1/2$ to 6 minutes, or until hot and just set, rearranging once or twice.

PENNSYLVANIA QUILTING, SHOOFLY PIE AND APPLE PAN DOWDY

Minnie Schlegel was raised on a dairy farm "in the heart of Pennsylvania Dutch country." (Though of German origin, her family is not Mennonite.) She's been active in the Reading Fair in eastern Pennsylvania since the 1950s, taking prizes for apple pie and quilting. "I won the Farm Maid contest in 1958," she recalls. "You had to milk a cow, make a dress and make an apple pie. Be hard to hold that contest nowadays. Cows aren't used to hand milking, and they'd probably kick you or step in the bucket."

Quilting has always been a keen interest. "My family did a lot of quilting. When I was growing up neighbors got together; a family'd put a quilt base on a frame. You'd put on the backing, then the batting, piece the top, and quilt through three layers. Ten or twelve can work on a quilt. Refreshments like lemonade and homemade ice cream went along with the quilting. The men turned the ice cream crank while the women quilted. Later a hot meal was served. If there were women present who did not quilt, they would thread needles and help with the refreshments.

The Quilting Party.

"Now I quilt with the Fleetwood Grange. We're making quilts bigger today, for the bigger beds people have. We quilt at the Reading Fair, and at the fair we raffle off one of the quilts we've made during the year. In our part of the country, the Dutch and the 'English,' as they call everyone else, quilt the same patterns, handed down through the generations. 'Dresden Plate' is a popular one. That's circles on squares, and a circle can be made up of thirteen to thirty patches, depending on the variation you are doing. 'Pieced Star' is another pretty pattern, as is patchwork 'Double Wedding Ring' and 'Trip Around the World.'"

An excellent cook (Minnie has been preparing the meats—roast beef, turkey and ham—for the Fleetwood Grange banquets since 1957), she gives her recipes for two delectable Pennsylvania Dutch dishes. Dinah Shore made them famous in a 1940s song, "Shoofly Pie and Apple Pan Dowdy," by Sammy Gallop and Guy Wood.

SHOOFLY PIE

MAKES 1 (9-INCH) PIE

¾ cup plus 1 tablespoon all-purpose flour

1 cup firmly packed light brown sugar

¾ teaspoon baking powder

Pinch of salt

5 tablespoons butter

⅓ cup molasses

1 egg

¾ teaspoon baking soda

¾ cup boiling water

1 (9-inch) unbaked pie shell

- Preheat the oven to 350 degrees.
1. Prepare the crumb topping: In a bowl, blend 3/4 cup of the flour, 3/4 cup of the brown sugar, the baking powder, salt and butter. Rub them together until the mixture resembles coarse crumbs.
2. Make the filling: In a bowl, mix together the remaining 1/4 cup of light brown sugar, the molasses, the egg, the remaining tablespoon of flour and the baking soda. Pour on the boiling water. Mix well.
3. Pour the filling into the unbaked pie shell and sprinkle the crumb topping evenly over it. Bake in the center of the oven for 35 minutes, or until the filling is set and does not quiver when the pan is shaken. (Do not overbake or the pie will be dry.)

APPLE PAN DOWDY

MAKES 1 PAN DOWDY

1 cup brown sugar, firmly packed	½ teaspoon ground cinnamon
¼ cup all-purpose flour	1 teaspoon vanilla extract
¼ teaspoon salt	6 cups peeled, cored, and sliced apples
1½ teaspoons cider vinegar	1 tablespoon butter, melted
1 cup water	

TOPPING:

1 cup all-purpose flour	2½ teaspoons butter
½ teaspoon salt	¾ cup milk
2 teaspoons baking powder	

- Preheat the oven to 375 degrees.
1. Put the brown sugar, the 1/4 cup flour, and the 1/4 teaspoon salt in a 1-quart nonaluminum saucepan. Add the vinegar and water and cook over moderate heat, stirring constantly , until thick. Remove from the heat and let stand until you can comfortably touch the mixture. When cool, stir the cinnamon, vanilla and butter into the spiced syrup.
2. Generously butter a 7 × 11-inch (6-cup) glass baking dish or its equivalent. Arrange the apple slices in it. Pour the spiced syrup over the apples.
3. Make the topping. Whisk the flour, salt and baking powder together. Cut in the butter until the mixture resembles coarse meal, then stir in the milk just until combined. Drop by tablespoonfuls over the apples. Bake for 45 minutes, or until bubbling.

HOOKSTOWN FAIR
Hookstown, Pennsylvania

FAIR DATE: LATE AUGUST, FOR FIVE DAYS, ENDING THE SATURDAY BEFORE THE LABOR DAY WEEKEND

GAZETTEER: *The Hookstown fairgrounds are located in Western Pennsylvania, in the south-western tip of Beaver County, very near to both West Virginia and Ohio. For the fairgrounds, take PA 168, one-half mile north of the intersection of PA 168 and U.S. 30.*

"We're a rural community surrounded by small cities, and we're not far from Pittsburgh, so we draw visitors from a large area," says Marie Elliott, whose husband, Chester, is chairman of the fair board. The fair is owned and operated by the Hookstown Grange. Back in 1947, there was a need to build a Grange Hall, and a need to find the money, too. A fair was the answer. The first fair was a success, and the Grange bought some property. "Now," says Marie, "we have a fine Hall and a first-class, seventy-acre fairground with all the facilities you could think of.

"There's lots of opportunity for the 4-H kids to participate. They can show and sell their steers, pigs, lambs and veal calves, and they can show their horses. As an agricultural fair, we want to educate you as well as entertain you. You can watch local farmers working with draft horses, just as the farmers did before the days of huge diesel tractors and harvesters. You get an idea of just how hard it was to till the land until very recently.

"The fair cafeteria is operated by the Mill Creek United Presbyterian Church, but the recipes were all originated by Grange women. We all work together to make pies—mass production! These recipes are used at Grange events all year. Some of our best cooks are so busy at fair time preparing food for the cafeteria that they can't take the time to enter contests as much as they'd like."

"This is a big area for car shows," says Michael Elliott, Marie's son and chairman of the Hookstown Fair Antique Classic Car Show. "People like to drive and talk; owners take pride in their cars, and like to show them to others. They come in Fords: Mustangs, Falcons, T-Birds and Torinos. There's Chevy Corvettes, and we had Belairs, Chevelles and Camaros. I'm a buff myself; I have a '67 Dodge Charger.

1937 Plymouth coupe.

Before the fair we send around a tabloid, asking for antique car buffs to come. We have judging in categories, such as cars made in the years 1935 to 1955, and '55 to '65. Most of the cars are restored. All cars must be pre-1974. Exhibitors like to get a prize or a ribbon; they're not interested in money. We have an impressive set-up, and we've learned that half the fairgoers come to look at the cars."

HOOKSTOWN GRANGE COCONUT CREAM PIE

"No one person is singled out as a master cook at our Grange," says Marie Elliott, "because ours is a group effort using recipes that have been created and modified through years of experience. According to a random vocal survey, the food most often named 'Our Best' was Coconut Cream Pie. We've modified our quantity recipe so that home cooks can use it."

MAKES 1 PIE

THE FILLING:

½ cup sugar	2½ cups milk
6 tablespoons cornstarch mixed with ½ cup milk	1 teaspoon vanilla extract
	1 tablespoon butter
3 egg yolks	½ cup shredded sweetened coconut
½ teaspoon salt	1 baked 9-inch pie shell

THE MERINGUE:

3 egg whites	5 tablespoons shredded sweetened coconut
6 tablespoons sugar	

1. Make the filling: Combine the sugar, cornstarch mixture, egg yolks and salt in a saucepan. Add the milk, stir, and bring to the boil. Immediately lower the heat and cook, stirring constantly, until the mixture thickens.
2. Off the heat, blend in the vanilla, butter and coconut. Let the custard cool for about fifteen minutes, then pour it into the pie shell. While it cools, preheat the oven to 350 degrees.
3. Make the meringue: In a large bowl, beat the egg whites with a whisk or an electric mixer until very frothy, gradually beat in the sugar, and continue beating at high speed until the whites stand in unwavering peaks. Spoon the meringue on the warm pie filling, making decorative swirls with the back of a spoon. Spread it to cover all the filling, sealing the edges to anchor the meringue so that it does not shrink. Sprinkle on the coconut. Bake for 5 to 10 minutes, until lightly browned. Chill thoroughly before serving.

THE PENNSYLVANIA FARM SHOW
Harrisburg, Pennsylvania

FAIR DATE: EARLY JANUARY, 6 DAYS, SATURDAY TO THURSDAY

GAZETTEER: *The arena is located at Cameron and Maclay Streets, just off Exit 32 of I-81.*

In central Pennsylvania, the snow and cold of early January make what the locals call "Farm Show weather." This event is so much a part of life here that it has entered into the language. Just about everyone, rural and urban, tries to make it to the huge complex housing the show. What began as a modest gathering of farmers in 1917 has become the largest free indoor agricultural exposition in the country, showcasing the state's largest industry—agriculture.

On Saturday in the large arena, huge and placid draft horses stand patiently for conformation judging; later, the Pennsylvania High School Rodeo competitors show considerable bustle as they rope and tie calves, and race around the barrels. Elsewhere, prize poultry squawk, swine squeal, and city kids watch as cows are milked. In the vast Family Living area, the Society for the Responsible Use of Animals, sponsored by the Dairy Council and Pennsylvania State University, brings awareness of proper care to both farmers and the public. Endless rows of canning jars catch the light, and people crowd the seats set up before the judging stand where cakes, cookies and apple pie will be judged.

Contestants represent their own county fairs, and a blue ribbon is proof that you've been judged the absolute best in the state of Pennsylvania.

SHE TAKES THE CAKE IN PENNSYLVANIA

Sandy Johnson rode from her home in Tioga, five miles from the New York border, to the Farm Show, balancing her cake on her lap. Both baker and cake survived in good enough shape to take first prize in the Greatest Chocolate Cake contest, cosponsored by the Pennsylvania State Association of County Fairs and Hershey Foods.

In the original competition of ninety-four county and local fairs, one thousand one hundred and eighteen cakes were up for honors. Division winners were eligible to compete at the Farm Show; sixty did.

Sandy and her husband, with five kids ranging from grade school to out-in-the-world, live on a four-hundred-acre dairy farm. She's a 4-H leader, and she bakes and decorates cakes for anyone who wants one—"It's no big thing," she says. This cake won first prize at the Tioga County Fair.

SANDY JOHNSON'S WINNING COCOA CAKE

Sandy says to use your favorite buttercream frosting.

MAKES ONE 3-LAYER CAKE

½ pound (2 sticks) unsalted butter, at room temperature

3 cups lightly packed light brown sugar

4 large eggs

2 teaspoons vanilla extract

¾ cup Hershey's cocoa

3 teaspoons baking soda

½ teaspoon salt

3 cups *minus 6 tablespoons* all-purpose flour

1⅓ cups sour cream

1⅓ cups boiling water

- Have ready three 9-inch cake pans, sprayed with vegetable spray. Preheat the oven to 350 degrees.

1. If possible, use a heavy-duty standing electric mixer. In the large bowl, cream the butter, beating it until light and fluffy. Add the brown sugar slowly, beating until the mixture is light. Add the eggs, one at a time, beating well after each addition. Beat in the cocoa, soda and salt. Add the flour, alternating with the sour cream. Beat on low speed until blended. Add the boiling water and stir until blended.

2. Divide the batter among the pans. Bake in the center of the oven for 35 minutes, or until a toothpick inserted in the center of the cake comes out clean. Cool the pans on racks for 10 minutes. Turn out the cakes and let cool completely on racks.

3. When the layers have cooled, ice the cake.

Chocolate cakes up for judging.

QUEEN OF THE COUNTY FAIRS

Each year, a poised and civic-minded young woman is chosen by the Pennsylvania State Association of County Fairs to represent that body. Cheryl Anne Muraski served in 1992, when she was a high school senior. "I had my queen duties and so many school activities I didn't see how I could earn money for college," Cheryl Anne says, "til I came up with the idea of baking cakes and selling them. Mother says I've been happy in the kitchen since she put me on the counter in my baby carrier to watch her cook." Cheryl Anne plans a career in the culinary field.

Queen Cheryl Anne Muraski.

SHRIMP WITH LEMON NOODLES

"This dish is so easy," says Cheryl Anne, "but it's fancy enough for a small dinner party."

SERVES 4 TO 6

1 2 pounds medium shrimp	Juice of 1 lemon
1 pound narrow egg noodles	1 teaspoon grated lemon peel
5 tablespoons plus 1 teaspoon butter	¼ cup minced parsley
1½ cups sour cream	

- Preheat the oven to 375 degrees.
1. Peel the shrimp, devein them, and cook them in lightly salted boiling water until they turn pink and lose their translucency, about 1 minute. Drain and set aside.
2. Cook the noodles according to package directions. Drain and place in a baking dish. Melt the butter and pour over the noodles. Stir in the sour cream, lemon juice and lemon peel. Toss in the shrimp. Bake 15 to 20 minutes, until hot and bubbling.

CHERYL ANNE MURASKI'S BEST-EVER SWEET POTATOES

"Not just for Thanksgiving!" says Cheryl Anne.

SERVES 6

½ teaspoon ground cinnamon

½ cup all-purpose flour

½ cup firmly packed brown sugar

½ cup quick-cooking oatmeal

½ cup chopped pecans

8 tablespoons (1 stick) butter,
   at room temperature

2 (40-ounce) cans sweet potatoes,
   drained

2 cups fresh cranberries

2 cups peeled, cored and sliced
   cooking apples

- Preheat the oven to 350 degrees.
1. Mix the cinnamon, flour, brown sugar, oatmeal, pecans and butter together until the mixture is crumbly.
2. Butter a 2-quart casserole. Layer it with half the sweet potatoes. Top with a layer of half the cranberries, and top that with half the apple slices. Sprinkle evenly with half the crumb mixture. Repeat the layers, and top with the remaining crumb mixture.
3. Bake uncovered for 35 minutes, or until the top is lightly browned.

THE GREAT 1890 COUNTRY FAIR

Dover, Delaware

FAIR DATE: OCTOBER, EVERY OTHER YEAR, FOR 1 DAY

GAZETTEER: *The Country Fair is a creation of the Delaware Agricultural Museum and Village, 866 North DuPont Highway, in Dover.*

Step right up, folks! And step back in time, to participate in a country fair just as it might have been one hundred years ago. Watch a pugilistic demonstration, as "Gentleman Jim" engages in the manly art against "John L. Sullivan." Indulge in Cracker Jack, gingerbread and lemonade (hot dogs won't enter the scene for at least another decade), as you listen to a lecture on the Evils of Suffrage for Women. The right to vote, according to the speaker, will not only destroy the family, but will lead to women who are "large-handed, big-footed, flat-chested, and thin-lipped."

"The fair is living history," says Mary Kopco, museum curator. "The museum has an exhibit hall devoted to technology and its effect on farming and work. Outside, in our village and farmstead, time stops in 1890. We have buildings here from all over the state, and restored to that time. It's the perfect setting for a fair."

A fair event that confuses many people is jousting; knights on horseback in medieval costumes, spearing rings on their lances. "How can this be 1890?" they wonder. But Victorians were fascinated by the Middle Ages, and tilting, or running at the rings, was a big fair sport for them.

Politicians harangue the crowd about issues of the day. (Not *our* day, but back then.) And these are genuine, modern politicians, who get a kick out of impersonating long-gone legislators. You can visit the House of Wonders and be amazed—or bamboozled—

by the bottle of performing fleas, or the thought reader. But watch out for the hussies and the fakirs!

"Company founders often introduced new products at fairs during the 1890s," says Mary. "'Henry J. Heinz' will give you a pickle pin and a sample of his new baked beans; 'Charles Elmer Hires' will offer you his temperance drink, Hires Root Beer; and 'Walter Baker' will give you a taste of his chocolate."

1890 FAIR RECIPES

"These recipes might have been prepared for judges to test or visitors to enjoy at a country fair during the 1890s," says curator Mary Kopco. "They were taken from a 1941 publication of the Delaware Home Economics Association, here in Dover."

SASSAFRAS TEA by Carry Lecates Kleinheimm
In the spring of the year, slice off bark of sassafras root (1/2 cup bark). Let simmer in 2 quarts of water until water is well-colored. Make fresh with each meal. Add 1 teaspoon sugar to cup, and cream if desired.

PEACH CORDIAL by Mrs. Henry Moor
Take the finest peaches, pare and pack them as closely as you can in a large stone jar. Then pour over them the finest spirit, let them stand 48 hours, and have it strained and sweetened to your taste.

SAD DUMPLINGS by Mrs. C. R. Donaho
Sift flour and salt together, add enough boiling water to make dough. Roll out 1/4 inch thick and cut in 1-inch cubes. Good with stewed chicken and excellent with green peas and butter.

WHEAT COFFEE by Mrs. Gardner Ellis
Parch the wheat, grind in coffee mill. Make the same as coffee quickly, and serve unsweetened with cream while hot. Delicious!

DELAWARE STATE FAIR

Harrington, Delaware

FAIR DATE: LATE JULY, 10 DAYS

GAZETTEER: *Harrington is 65 miles south of Wilmington. This area is part of the Delmarva (DELaware, MAryland, VirginiA) peninsula, one of the major chicken-raising regions in the United States. The fair is right off Route 13, the main route though Delaware.*

Delaware's State fair is a splashy delight, with harness racing, agricultural displays, stock car racing, booming midway and entertainment from big-time singing acts to Porky, "the world's largest pig," eleven hundred pounds from snout to curly tail.

Many fairs used to depend on railroads, to carry the farm animals for exhibit and to bring eager city billies for a day of bucolic fun. In fact, a number of fairgrounds were built near the local railroad station, for just those reasons. Now, of course, agricultural exhibitors mostly come by truck and trailer, and visitors must now drive their own cars.

Douglas R. Andrews, president of the Delmarva Rail Passengers Association, a private group with one hundred members, thought that was a shame. "I can remember taking the train to the fair," he recalls. Regular passenger rail service in Delaware came to a halt in 1965. But in 1992, after months of work with other agencies, including Amtrak and Conrail, Andrews' group arranged a one-day excursion from Claymont, with passenger pickups in Wilmington, down to Harrington. Fairgoers could disembark and spend five and one half hours at the fair, or go further south to Millboro, turn around, come back to the fair and spend a shorter time. The five air-conditioned cars, one a restored Pennsylvania Railroad private car, then took the passengers home. All five hundred places on the cars were filled with happy, enthusiastic riders, combining a scenic trip with a grand fair day, and no parking hassles. Andrews hopes this success will reestablish north-south passenger rail service in Delaware, and not just for the fair.

JEAN MILLER'S CHICKEN POTPIE

"Being as how we're in the midst of chickens here in Delmarva, it's patriotic to use them as much as possible," says Jean. She retired from the Department of Defense after 31 years, and now works part-time as ticket office manager for the Delaware State Fair. Jean is active in community affairs, collects black amethyst glassware, and loves to cook. Jean and a friend have an on-going fudge competition at the fair: "One year Jane will win the blue ribbon and I'll get the red, and the next year it's reversed. The funny thing is that we both use the same recipe that we got off a jar of marshmallow cream."

SERVES 4

THE PIE:

⅓ cup diced celery

⅓ cup diced onion

1 cup sliced carrots

4 cups chicken broth

1 tablespoon cornstarch mixed with

¼ cup water

1 cup cooked fresh peas or defrosted frozen

2 cups cooked chicken, cut into large chunks

Salt and freshly ground pepper to taste

1 teaspoon dried basil or thyme, crumbled

THE BISCUITS:

2 cups all-purpose flour

1 teaspoon salt

1 tablespoon baking powder

6 tablespoons vegetable shortening

⅔ cup milk

• Preheat the oven to 350 degrees.

1. Gently cook the celery, onion and carrots in the chicken broth until just tender. Add the cornstarch mixture, stir, and cook a few minutes over moderate heat until thickened. Add the peas, chicken and basil. Put the mixture in a wide casserole that just fits it.

2. Make the biscuits. Mix the flour, salt and baking powder. Cut in the shortening with a pastry cutter, or use your fingers to rub the flour and fat together until the mixture resembles coarse meal. Pour in the milk and stir with a fork just until the dough holds together. Knead twelve times, about 30 seconds. Roll out on a lightly floured work surface to a circle 1/2 inch thick. Cut the biscuits with a biscuit cutter and place on top of the pie. Bake in the center of the oven about 30 minutes, until the broth is bubbling and the biscuits are a delicate brown.

KENT COUNTY FAIR

Chestertown, Maryland

FAIR DATE: THIRD WEEKEND IN JULY, 3 DAYS

GAZETTEER: *The fairgrounds are 1 mile from Chesapeake Bay, on Maryland's Eastern Shore. Take Route 20 West from Chestertown, turn right on Route 21 and look for signs to Kent Ag Center. Open rural countryside, much like the Midwest. Population of farmers and watermen; goose hunting in fall.*

"Ours is a 4-H fair, not cutthroat, and any kid can feel good when they leave," says Susan Debnam, who chaired the first fair some fifteen years ago and is in charge of all food. "We have no carnival, and we cook all the food for the fair ourselves, no concessions, except for Future Farmers of America selling ice cream. We have a wonderful old military kitchen; that's because our fairground was an old NIKE base that the county leases to us. You know what wonderful crabs come from Chesapeake Bay. Kent County Fair features crab cakes, and we're getting famous for them!"

A 4-H exhibitor since the third grade, Sara Morris lives on a four-hundred-and-sixty-five-acre dairy farm. She likes to cook, and in summer usually makes dinner for the family. As an exhibitor, she's done better than well in a number of projects: "I've won Senior and Grand Championship in food preservation and fashion modeling, and Junior Championship in clothing construction.

"You work a full year for the Kent County Fair. You do your record book, which is about how your project developed and your skills grew, in the winter; and when you show your project in the summer, you are judged on your book. The fair is the highlight for us.

Winner and friends.

"You don't know what it's like to live here. We don't have McDonald's, or The Gap, and we can't get cable TV. Last year when we got a new shopping center, they had to bring in the state police to control the crowds! We live on Mary Morris Road, named after my great-grandmother. We've been here for a hundred and fifty or so years. My brother wants to take over the farm, but I want to move to the city."

SARA MORRIS' ZUCCHINI SQUASH RELISH

"I learned to can food helping Mom," says Sara. "We have a big garden and orchard, and we put away a lot of things. This relish is excellent on hot dogs—and it uses up a lot of zucchini!" Sara's relish won the Senior 4-H Food Preservation award for 4-H'ers over fourteen.

MAKES ABOUT 8 1/2 PINTS

10 cups shredded unpeeled zucchini	5 teaspoons pickling salt
4 cups shredded onions	2½ cups cider vinegar
3 large green peppers, cored, ribs removed, and shredded	5 cups sugar
	2 teaspoons celery seed
1 large red pepper, cored, ribs removed, and shredded	1 teaspoon turmeric
	½ cup cornstarch

1. Place the zucchini, onions, and peppers in a large nonaluminum bowl. Mix the salt in well and let stand overnight. Next day, rinse well under running cool water and drain thoroughly.
2. Make the syrup. In a very large pot combine the vinegar, sugar, celery seed and turmeric. Bring to the boil. Add the drained vegetables and bring to the boil again; lower the heat and simmer, partially covered, for 15 minutes.
3. Mix the cornstarch with enough water to make a smooth paste. Add to the pot and stir well; cook a few minutes until the liquid thickens slightly and the cornstarch has lost any raw taste.
4. Spoon into hot, sterilized pint or half-pint jars. Process in a boiling water bath for 10 minutes.

KENT COUNTY FAIR MARYLAND CRAB CAKES

"This is the recipe we've used for years," says Susan Debnam. "We serve the crab cakes on hamburger buns, with lettuce, tomato, and ketchup on the side."

MAKES 6 CAKES

1 pound fresh Maryland crabmeat	¼ teaspoon freshly ground pepper
1 cup Italian seasoned bread crumbs	1 teaspoon Worcestershire sauce
1 large egg	1 teaspoon dry mustard
About ¼ cup mayonnaise	1 teaspoon Old Bay Seasoning (optional)
½ teaspoon salt	Vegetable oil for frying (see Note)

1. Pick over the crabmeat, removing all cartilage.
2. In a large bowl, mix the bread crumbs, egg, mayonnaise and seasonings. Add the crabmeat and mix gently but thoroughly. If the mixture seems too dry, add a little more mayonnaise. Cover with plastic wrap and refrigerate overnight to let the flavors blend.
3. Next day, shape the mixture into 6 cakes. Heat 1/2 inch of oil in a heavy skillet. Cook the cakes until browned, about 5 minutes on each side.

NOTE: If desired, crab cakes may be deep-fried in oil heated to 375 degrees (use a deep-fry thermometer) for 2 to 3 minutes, or until browned.

ELIZABETH MORRIS' OYSTER BRIE HORS D'OEUVRE

Sara's mother, Elizabeth, has acted as consultant to the Maryland Office of Seafood Marketing, developing recipes. "Oysters are never served at the Kent County Fair," she says, "because it is held in July, and we serve oysters only in months that have an 'R.' But oysters, like crabs, abound in Chesapeake Bay."

MAKES 18

18 Maryland oysters, in the shell	2 tablespoons pimiento, finely chopped
3 ounces ripe Brie cheese, chilled	Lemon wedges

- Preheat the oven to 450 degrees.
1. Open the oysters, leaving the meat in the deeper half of the shell. Arrange on a baking sheet or in pie tins containing a bed of rock or kosher salt. Slice the chilled Brie very thin, and lay a slice of cheese on each oyster. Top each with a little pimiento.
2. Bake for about 10 minutes, or until the cheese is melted and the oysters are plump. Serve with lemon wedges.

THE VISUAL ORANGE SMOOTHIE

Sara Morris of Chestertown, Maryland, made her first Visual Presentation as a first-year 4-H Club member in 1984. Her mother, Elizabeth, recalls Sara's big moment: a Visual Presentation is defined as an informal talk and demonstration that teaches the audience how to perform a task. It takes a lot of practice and many run-throughs until the 4-H'er knows the routine. Sara really practiced making her 'Smoothies.' She won both Blue and Champion ribbons, first at the county level and then at the Maryland State Fair. Quite a feat for an eight-year-old!"

SARA'S SMOOTHIES

SERVES 6 TO 8

1½ cups skim milk
1 (6-ounce) can orange juice concentrate, thawed
1½ cups water
1½ teaspoons vanilla extract (optional)

- Combine all ingredients in a large glass bowl. Beat with a wire whisk or hand-held beater until very frothy. Serve at once.

Young Sara Morris makes her Orange Smoothie.

MONTGOMERY COUNTY AGRICULTURAL FAIR

Gaithersburg, Maryland

FAIR DATE: 9 DAYS, ENDING 1 WEEK BEFORE LABOR DAY

GAZETTEER: *Gaithersburg is about 25 miles north of Washington, D.C., on Highway 270.*

Over seven hundred thousand people now live in Montgomery County, but in 1846, when concerned citizens organized a gathering called the Rockville Fair, the population had dwindled alarmingly. Tobacco had so depleted the soil that farmers were moving to fertile farmland to the west. The "agricultural exhibition," as it was then called, was held to inform those who remained about better agricultural techniques and equipment. And of course, as always, the fair served a social purpose. But by 1933, the Rockville Fair had fallen on hard times and closed.

Now there is a new fair, without harness racing or betting, but with agricultural displays to inform a city-raised fairgoing public that has probably never seen a cow being milked, or beef outside a meat case. "Not long ago here in the country you saw one dairy barn after another," says Roscoe Whipp, secretary of the fair. "Washington is getting too close! The fair really started around 1945, because the local farm kids had no place to show their animals. The 4-H was the purpose behind the fair. We're big on the home arts and crafts, too. Now we have about forty buildings.

"All kids under twelve get in free. We try to appeal to kids. We bring very ill

Old Rockville Fairgrounds, 1920s.

children to the fair, and some of the fair activities go to the children's hospital in the town. Once we had a purple cow, dyed with gentian violet. I'd like to bring that back! We like to feature dairy cattle breeds, which is how we started with our huge wheels of cheddar cheese."

"I'm what they call a 'down county' girl," says Helen Hubbard, superintendent of Home Arts, "which means I don't live in the farm part of the county. But we started coming when I entered my mother's crocheted afghan and it won first prize. She got interested, and demonstrated crocheting until she was ninety. Now my husband Frank and I volunteer, and actually we all start to work on the next fair just after the current one closes. My sons' wives are active in Home Arts, and my twelve-year-old granddaughter, Gwendolyn, makes toasted cheese sandwiches. I'd say we have about three thousand pounds of cheese, in huge wheels, and most of it goes.

"We're not a farm family, but seeing agriculture is important to us. That should be part of everyone's life.'

HELEN HUBBARD'S COLESLAW

"The people at my church, Glenmont United Methodist, call this my specialty. What makes it special is a lot of pepper, I think. Of course you can use less. And if there's people with false teeth, I eliminate the celery seeds that can get under their plates."

MAKES 6 TO 8 SERVINGS

1 medium-size green cabbage, about 2 pounds	¼ cup sugar
2 carrots, peeled	2 tablespoons cider vinegar
1 small onion, peeled	½ teaspoon salt
1 cup Hellmann's or Best Foods mayonnaise	2 teaspoons freshly ground pepper
	1 teaspoon celery seed (optional)

1. Shred the cabbage by hand, or shred or chop it in a food processor. The cabbage should be somewhat coarse. Chop the carrots and onion until fine.
2. Place the vegetables in a large bowl and mix in the rest of the ingredients. Refrigerate, covered, for at least 4 hours to let the flavors blend.

HOT MILK CAKE

"This cake won first prize in 1988," says Helen Hubbard. "It's very nice with fruit, or just a cup of tea. The hot milk seems to give it a lovely texture. Be sure and put the baking powder in last—that's important. Otherwise the cake doesn't rise properly."

This light, delicious cake could be drizzled with a confectioners' sugar or chocolate glaze, or served with sugar-sprinkled sliced strawberries. You could also add a teaspoon of orange or lemon juice and a teaspoon of grated orange or lemon peel to the batter.

MAKES 1 CAKE

1 cup milk	2 cups sifted all-purpose flour
4 tablespoons (½ stick) butter	1 tablespoon vanilla extract
4 large eggs	¼ teaspoon salt
2 cups sugar	2 teaspoons baking powder

- Preheat the oven to 350 degrees. Thoroughly grease a 9- or 10-inch tube pan.
1. Heat the milk and butter together in a saucepan until the butter is melted and the mixture lukewarm. In a large bowl, beat the eggs until foamy. Add the sugar gradually, as you beat, and beat until light and fluffy. Add the milk-butter mixture. Add the flour and beat well, then beat in the vanilla and salt. Last of all, beat in the baking powder.
2. Pour and scrape the batter into the pan. Bake 45 minutes, or until a toothpick inserted in the center comes out clean. Cool on a rack.

Charter Jersey Club members in 1982. (Club founded in 1924.)

ROCKINGHAM COUNTY FAIR

Harrisonburg, Virginia

FAIR DATE: THIRD WEEK IN AUGUST, FOR 7 DAYS, SUNDAY TO SATURDAY

GAZETTEER: *The fairgrounds are in the beautiful Shenandoah Valley, on U.S. Route 11, 1/2 mile south of the Harrisonburg city limits. On I-81, take exit 240 or 243. Nearby are New Market and other Civil War battlefields.*

"I've heard from out-of-state visitors that our fair is heavily oriented toward agriculture in proportion to rides and games," says Dennis Cupp, the fair's general manager. "At some fairs you must look hard to find agricultural exhibits and the handiwork that the ladies have done. Ours is the number-one agricultural county in the state of Virginia. We're number one in poultry in the state, and among the top ten producers in the nation.

"We reflect our ethnic heritage; we're German, and some Scotch-Irish, and we're tight with the money—not much gambling. The Blue Ridge Mountains separated us from the English settlers in Richmond and the Tidewater. The Shenandoah Valley has always been family farming territory. The farms of the early 1700s were subsistence operations, and one of the first cash products of the frontier was whiskey: easy to haul on horseback, and consumed by a ready market. Agriculture here has changed since those days!

"We have Mennonites around here too, and you must be careful not to run over their horse-drawn buggies. The Old Order Mennonites won't go to the fair, but a lady who helps me out says she can spot the teens who change clothes and sneak in. They let down their long hair and it flies in the wind on the midway rides."

Basting the chicken barbecue.

MY FAVORITE CORNBREAD

Norma Cupp, whose husband is a relative of fair manager Dennis Cupp, has been judging baked goods at Rockingham for twenty years. "I first made this cornbread when I was milking two Guernsey cows and had lots of cream and butter," Norma says. "The cornbread gets tall and light."

MAKES 1 SQUARE CORNBREAD

¾ cup yellow cornmeal

1¼ cups sifted flour

1 tablespoon baking powder

½ teaspoon salt

¼ cup sugar

2 eggs, beaten

1 cup heavy cream

• Preheat the oven to 425 degrees.

1. Sift or whisk the dry ingredients together. Add the eggs and cream. Beat hard until smooth. Pour into a greased 9 × 9-inch pan. Bake about 25 minutes, until firm.

Delighted winner Robert Simmons, Jr., leading a Hereford bull.

JACKSON COUNTY JUNIOR FAIR

Cottageville, West Virginia

FAIR DATE: 5 DAYS, LAST WEEK IN JULY

GAZETTEER: *The fairgrounds lie in the Ohio Valley, on the eastern border of the Ohio River, 35 miles north of Charleston. Take Route 33, six miles west of I-77, and get off at the Ripley exit. This is beef-raising country of rolling hills.*

The fair began in 1957, to fill a need: a place for 4-H, Future Farmers of America (FFA) and Future Homemakers of America (FHA) to display their talents. No adult exhibitors! That two-day affair has grown into a five-day fair, with some five hundred youngsters participating.

Along with serious competition, showing such livestock as calves, goats, rabbits and sheep, there's the lighter side of a youth fair: the Greased Pole Climb, contenders shinnying up a twenty-foot pole; Looking for Money in a Haystack, kids scrambling for —and keeping—nickels, dimes and quarters; and the Tobacco Spitting Contest, with watermelon seeds replacing the nicotinic plug. There's also a pet show.

Bethany Waskey is an enthusiastic competitor. "I've been going to the fair since I was born," she says. At the age of fourteen, Bethany, a 4-H'er, won Best of Fair for her cookies, and entered the Small Pets event. "I have a little black terrier with a curly tail named Midnight, and I taught her to do tricks. Well, I taught her to sit—that's about all she can do." Bethany next plans to enter her African pigmy goat, Jessica, in Small Pets, and younger brother Robert will enter Mary. "We got the goats when they were two days old and their mother had died," Bethany says. "We bottle fed them, and they are really tame little pets."

Searching for money in a haystack.

Greased pole climb.

BETHANY'S BLUE RIBBON
CHOCOLATE CHIP COOKIES

"I just played around with recipes until I got the taste I like," Bethany says. "I cook lots of stuff. Bread, and pizza from scratch." The recipe may be doubled.

MAKES 2 DOZEN 3-INCH COOKIES

½ cup granulated sugar

¼ cup light brown sugar

½ cup Butter Crisco

1 large egg

1 teaspoon vanilla extract

1 cup sifted flour

¾ teaspoon salt

½ teaspoon baking soda

1 6-ounce package miniature semisweet
chocolate chips

- Preheat the oven to 375 degrees.
1. Combine the sugars, the Crisco, egg and vanilla. Beat together with an electric mixer or wooden spoon until light and fluffy.
2. Stir in the flour mixed with the salt and baking soda. Fold in the chips. Drop by the spoonful onto lightly greased cookie sheets, 2 inches apart. Bake for 10 to 12 minutes. Remove and let cool completely on racks.

Bethany Waskey.

THREE

THE SOUTH

Shelbyville

Kentucky • Lebanon

North Carolina

• Concord

Tennessee   • Gray
Lebanon

• Anderson

Hiawassee   **South Carolina**

Mississippi   **Alabama**   **Georgia**

Philadelphia   Dothan

Inverness
•
Eustis

Florida •Miami

*North Carolina, South Carolina,
Kentucky, Tennessee, Georgia, Florida,
Alabama, Mississippi*

CABARRUS COUNTY FAIR
Concord, North Carolina

FAIR DATE: SECOND WEEK IN SEPTEMBER, 6 DAYS

GAZETTEER: *Concord is about twenty miles east of Charlotte. The Charlotte Motor Speedway is ten miles away. For Concord, take Highway 29. Fairgrounds are at 789 Cabarrus Avenue West, on the edge of Concord.*

Dona, the baby Indian elephant.

"The first gold discovered in America was in Cabarrus County," says Harold McEachen, the fair's general manager. "In 1799, a twelve-year-old, Conrad Reed, discovered a seventeen-pound nugget. Reed's Gold Mine is now a state historic site. Scotch-Irish and German settlers came here well before the revolution, and most people hereabouts are their descendants.

"We have an old-timey fair just like they held seventy years ago; we haven't modernized a lot. We're friendly. Many come out here every day just to watch people. We're big on poultry; we've got classes for American, Asiatic, Mediterranean, English and Polish breeds as well as game birds and fancy pigeons. The Piedmont Bantam Association is pretty strong here. We have just started donkey and mule classes. That's really getting popular; folks get a kick out of them.

"Our entertainment is free, and we like to say 'we knew them when.' Hank Williams, Jr., was here when he was just nineteen, and before his show all the lights blew out. Some mighty quick work fixed the transformer, and the show went on. Fourteen-year-old Tanya Tucker stayed overnight one year, in my daughter's room. Now she's a famous country singer."

DORIS ROGERS, SCRATCH COOK

Mrs. Rogers is the Home Economics Agent of the North Carolina Extension Service, teaching people to improve or maintain a good quality of life. She's in charge of twenty-four Extension Homemakers Clubs, and she trains volunteer leaders. "I like to say I can cook from scratch quicker than I can open a package. I come from the casserole era, but that's too many calories, so we are trying to find ways to cook that are simple again. We are adapting old recipes to make them lighter.

"I've been with the fair over twenty-seven years," she says. "I'm in charge of educational services. I'm the seventh of nine children, and my father was a Baptist minister; we lived just over the county line, in Charlotte. My mother loved to cook, and I grew up on her good baking. I love fried chicken better that anything, and I never got enough; we kids only got one piece. So one time I went out, caught a chicken, and chopped off the head. The headless chicken started to run, but I caught it, plucked it like I'd seen my mother do, cut it up and fried it, and ate the whole thing!"

DORIS' FRIED CHICKEN

"My daughter, Nancy White, cooks chicken this way too," says Doris. Nancy is married to a minister.

SERVES 6 TO 8

1 (3½ pound) chicken	2 tablespoons butter
½ cup flour	2 tablespoons sesame seeds
2 teaspoons paprika	3 tablespoons water
2 teaspoons salt	1 cup water from cooked potatoes, or
Lots of freshly ground pepper	any other vegetable
½ cup polyunsaturated oil, like canola	

1. Cut the chicken into serving pieces: you should have 2 legs, 2 thighs, and 2 wings. Then cut out the breast, or "pulley" bone; break off the breast cartilage and separate the two breast pieces with their ribs. Crack the back into two pieces.

2. Combine the flour, paprika, salt and pepper. Dust the chicken well, then shake off the excess. Save the excess flour mixture for gravy.

3. In a large heavy skillet, heat the oil and butter over high heat. Add the chicken and sprinkle with sesame seeds. Fry the chicken parts until brown on one side, then turn.

4. Add the 3 tablespoons of water to the skillet, lower the heat to moderate and cover the skillet. Cover and cook until tender, 20 to 30 minutes. Remove the chicken pieces to a plate.

5. Make the gravy. Over moderately high heat, add 2 tablespoons of the reserved flour and a little of the potato liquid. Scrape up the browned bits from the pan and stir constantly until the mixture is nicely browned. Add the rest of the potato water and cook until the gravy is thickened and brown. Add lots of pepper. Serve at once, with potatoes.

ANDERSON COUNTY FAIR

Anderson, South Carolina

FAIR DATE: SECOND WEEK AFTER LABOR DAY, FOR 9 DAYS

GAZETTEER: *Anderson is in the rolling foothills of the Blue Ridge Mountains in the Carolina Piedmont. The area is agricultural; it is the leading dairy county in the state, with soy beans a major crop. Clemson University is fifteen miles away. Coming from north or south, take I-85 and exit on 19A. Fairgrounds are about seven miles.*

The year 1920 marked the official opening of the Anderson County Fair, though perhaps the earliest fair in the area was held at Pendleton, in 1816. "Carnivals weren't a part of the activities of early fairs," says Hurley E. Badders, a fair historian. "The first to come to Anderson was in 1901, when the City Council paid the Sturis Company to set up by the courthouse." The stars of that event reverberate to this day for carnival buffs: they were the legendary Jo Jo the Dog Faced Boy, and the queen of the bewitching, twitching hips, the exotic dancer Little Egypt.

Today's fair boasts handsome buildings, a paved midway and a huge cattle barn. "At early fairs, livestock was limited to hogs and poultry, as registered beef and dairy cattle virtually did not exist in the region," says Mr. Badders. "Today, both beef and dairy cattle make up the largest showing in the state."

"Dairy beef and cattle and sheep can't all be shown at the same time, on account of Anderson County's so blessed with them," says John Gates, who calls himself an "old friend of the fair." He's worked in various capacities, among them as announcer, public relations man, and vice president. John recalls the great food of yesteryear—the Twenties, Thirties and later, when churches, lodges and private individuals worked at the fair: "Jim White had a fish house shaped like a pagoda, and he sold beer made in Brooklyn, New York. Ten cents a bottle. You could buy a fried fish sandwich for a quarter. The Good Neighbor Clubs from the cotton mills sold food, too. Most of the food was cooked at home and brought to the fair. At Geddings Lunch, set up in a tent, you could get pork chops, barbecue, collards, squash—the whole meal. Before integration, the blacks held their own fair here at the fair grounds, following the big fair. After integration, they came in with a sit-down meal that made the fair even better. Two pieces of chicken, turnip greens, string beans, cabbage, and always cornbread. About three dollars, and you couldn't beat it. Well, all that good food came to an end. The Health Department stopped it. Our state has the strictest food regulations in the United States. Now we have concessions. But the old-timers still reminisce!"

SHE KNOWS HOSPITALITY

"I feel that I've fed everybody in South Carolina at least once, and some of them twice," says Juanita Garrison. "We live in a hundred-year-old farmhouse that my husband's grandfather built after the War of Northern Aggression, as we like to say down here. My husband Ed is the fourth generation at Denver Downs Farm. Ours is a dairy farm, and the kids were all in 4-H. I was a member of the Extension Homemakers Club and a state officer, and I've also served as state president of the Master Farm Homemakers organization, for women whose families are involved in agriculture. I have competed in food preservation and baking and I'm also a nationally accredited flower show judge.

"My Ed was in the South Carolina state legislature for thirty years. I enjoy cooking for company; we've held teas, receptions, and every other year a barbecue on the Friday before the opening of the dove-shooting season. I have a huge U-shaped kitchen, and when I die, I say they can inter me right there, just like St. Peter buried in the Vatican."

The teapot refreshment stand, 1940.

JUANITA GARRISON'S WHITE MEAT CHICKEN SALAD

"Expand as you like, but keep the proportions correct. One half of the amount should be cooked chicken or turkey breast, cut into bite-sized pieces; the other half should be more or less equally divided among ingredients you select. You can use chopped celery, pecans, walnuts, almonds, grated onion, cucumber, pineapple tidbits—the list goes on! If you plan to make sandwiches, chop or grind the chicken fine. The salad is particularly pretty served in a large glass bowl, garnished with parsley. An excellent choice for luncheons and summer suppers."

SERVES 4, OR 8 AS PART OF A LARGE BUFFET

2 cups cooked, cut-up chicken or turkey breast
½ cup chopped pecans
1 cup finely chopped celery
½ cup white grapes, cut into quarters
Salt and freshly ground white pepper to taste
½ cup or more of mayonnaise, enough to bind

Mix all the ingredients together gently. Chill, covered with plastic wrap, until ready to serve.

DENVER DOWNS FARM CANDIED YAMS

"The yams or sweet potatoes should hold their shape but be tender and covered with a rich sticky syrup," Juanita says. "The baking soda helps keep the shape, I think. You can make them early in the day or even the day before if you omit the butter and refrigerate the cooked yams. About thirty minutes before serving, gently reheat them and add the butter."

SERVES 6 TO 8

2 to 3 large yams or sweet potatoes
1 tablespoon baking soda
¾ cup sugar
½ cup water
¼ teaspoon salt
½ orange, thinly sliced crosswise
3 tablespoons butter

1. Peel the yams. Slice or cut them into 1/4-inch strips. Put them in a bowl and cover with warm water; add the soda. Let stand 10 minutes, then drain.
2. Meanwhile, combine the sugar, water, salt and orange slices in a heavy saucepan and heat until the sugar is dissolved. Add the yams and cook, uncovered, over moderate heat until the syrup thickens and the potatoes are tender. This will take about 1 hour. Stir gently from time to time.
3. Just before serving, add the butter and stir the yams gently until the butter is melted and the yams warmed.

JUANITA'S GRITS

"In the South, grits are regularly eaten on the same plate with ham and eggs or bacon and eggs, and are an integral part of the meal, not just an accompaniment," says Juanita. "Usually butter is added to hot grits, unless the meal includes ham with lots of red-eye gravy. Most package instructions call for water only, but I prefer to use half water and half milk, or I sometimes use all milk."

SERVES 6

2 cups water and 2 cups milk, or 4 cups milk	½ teaspoon salt
	1 cup grits

1. Bring the liquid to a brisk boil, add the salt and slowly stir in the grits. Reduce the heat to very low. Cook for 20 minutes, covered, stirring from time to time.
2. When tender, remove from the heat and allow to stand, uncovered, a few minutes before serving.

MARION COUNTY COUNTRY HAM DAYS

Lebanon, Kentucky

FAIR DATE: LATE SEPTEMBER, FOR 2 DAYS

GAZETTEER: *Lebanon / Marion County, geographical center of the state of Kentucky. Settled in 1789, named for Revolutionary General Francis Marion, the "Swamp Fox." Agricultural land of rolling hills and pasture, producer of champion country hams, corn, cattle, soy and tobacco, and world-famous Maker's Mark bourbon whiskey. Within sixty miles: "My Old Kentucky Home"; Churchill Downs; Mammoth Cave National Park; Kentucky Horse Park, Lexington. Direction: Highway 68, downtown Lebanon.*

Almost everyone in Marion County, "The Heart of Kentucky," raised hogs in the old days; most everyone in the county farmed. Raising hogs and curing hams is now a specialized occupation, but the citizens know exactly how a country ham should taste—quality pork, dry-rubbed with salt and red and black peppers, that shrinks 20 to 30 percent during the twenty-one- to twenty-eight-day curing process. As the ham dries, the flavor intensifies, resulting in the rich, aromatic, deep complexity that distinguishes a Kentucky country ham.

Pride in the local product is what sparked the idea of the Ham Days some twenty years ago. Since that beginning, with a few borrowed stoves, a few tables and a few dozen hams sliced and cooked in Lebanon's Courthouse Square, the festival has grown to a rollicking two-day affair attracting more than fifty thousand visitors annually.

Ham, of course, is the focus, but when they aren't eating, visitors can watch the PIGasus Parade (a proud procession of porkers promenaded in supermarket baskets), cheer the hot air balloon race, hear the Marching Band Competition, huff and puff in the 10K Pokey Pig Run, pitch horseshoes or watch clog dancing. Both Hog and Husband Calling contests are popular, as is the Hot Pepper Eating Contest, though perhaps less so for the participants than the onlookers. Folks do better with ham.

The pinto bean cook and one of the fair's organizers is Dick Moraja, a civic leader and Lebanon's funeral director ("I was raised on a farm; farm boys like to be their own boss"). Dick remembers how it was not so very long ago: "The old staples were pintos, pork and corn bread. Salt pork and beans didn't need preserving," he says.

"When I was ten," continues Dick, who was born in 1941, "we used to raise about thirty hogs, sell some and slaughter a few for our own use. Six, eight neighbors brought over their hogs, too. We'd do this in the middle of November, when it was cold, but just before freezing. We didn't throw away much. The lard was rendered and used all year just like Crisco is now. We'd cut up hams and bacon, and clean and soak the small intestines for sausage casing. For sausages we'd use the bits of lean and fat left over, seasoned with sage, salt and pepper.

"To drive the sausage grinder, the men would jack up an old Model A and take off one back wheel. Take off the tire, and put a belt over the rim. The grinder was behind the car. They'd start 'er up, the engine turned the pulley, and sausage meat was ground. They used the belt to force the meat into the casings, too, then they'd hang them up and let them age."

Dick not only helps organize and cook for Country Ham Days, he's a competitor, too. "A bunch of us enter the contest for Best Ham. I use Ohio River salt, a wet salt for curing. It liquidizes, draws moisture out of the meat, and makes a brine. We let it stay down twenty-one days, then we hang it. It has to go through what we call the 'June sweats,' when the salt concentrates so the flavors develop. We eat the first ham about Thanksgiving, and ration them out over the year. I won the competition in 1991. Billy Joe, the mailman, won the year before that. Competition is pretty keen. Not too many understand curing."

Tall tale teller and antique engine mechanic Paul Cox, with Sam Roller.

MRS. MORAJA'S COUNTRY HAM STUFFED WITH GREENS

Preparing a ham with greens takes a commitment of time, but it is a triumph of Southern cooking. "Here's my mother's recipe," says Dick proudly. "When you slice the ham paper-thin, you have slivers of green stuffing in each slice." Dick recommends serving Country Ham with Greens accompanied by baked apples and mashed potatoes, followed by Jam Cake.

SERVES 12 TO 20

A 12- to 16-pound Kentucky country ham (or other country ham)

3 tablespoons butter

¼ cup chopped (green onions) scallions, white and green parts

1 pound fresh kale or mustard greens, or a combination, well washed and coarsely chopped

1 teaspoon crushed dried hot red pepper flakes

½ teaspoon salt

Black pepper to taste

1. Starting 2 days ahead, place the ham in a large pot and pour in cold water to cover by two inches. On the second day, change the water.
2. When ready to cook, replace the water with fresh. Bring to the boil, reduce the heat and simmer the ham for 1 hour. Transfer the ham to a cutting board. Rinse the pot and set it aside.
3. When the ham is cool enough to handle, cut off the rind and discard it. Cut off excess fat, leaving only a thin layer. Set aside.
4. Over moderate heat melt the butter in a large skillet. Add the scallions and cook, stirring, for 5 minutes until translucent. Add the chopped greens, pepper flakes, salt and pepper and cook, covered, over low heat for about 15 minutes or until tender. Stir occasionally. Set aside to cool.
5. Working lengthwise, make 6 to 8 deep 2-inch-long incisions in the ham, spaced about 2 inches apart. Hold the incisions open with a spoon, and stuff them evenly with the greens mixture.
6. Wrap the ham tightly in a double layer of cheesecloth or clean white cotton, such as an old pillow case. Tie the ends with kitchen cord—the ham must be tightly enclosed.
7. Return the ham to the pot, cover with cold water, bring to the boil, reduce the heat to low and partially cover the pot. Simmer 3 to 4 hours, or until the ham

is tender when pierced deeply with a skewer. The ham must remain covered with water; check from time to time and add boiling water if necessary.

8. Transfer the ham to a platter. Without removing the cloth, let it cool to room temperature. When cool, refrigerate for at least 12 hours. (Save the cooking water to cook greens.)

9. To serve, remove the cloth and carve paper-thin slices across the grain.

*Dr. James Kemp
judging country hams.*

KENTUCKY JAM CAKE

Jam cakes are a Kentucky tradition. Obviously, this is a grand cake for a grand occasion.

MAKES 1 LARGE (4-LAYER) CAKE

3½ cups all-purpose flour	2½ cups sugar
½ teaspoon salt	6 egg yolks, lightly beaten
1 teaspoon ground allspice	1 cup buttermilk
1 teaspoon freshly grated nutmeg	1½ teaspoons baking soda
1 teaspoon ground cloves	⅔ cups seedless raisins
1 cup blackberry jam	1 cup chopped pecans
1 cup strawberry jam	6 egg whites
1 cup peach preserves	Caramel icing (opposite page)
½ pound (2 sticks) unsalted butter, softened	

- Preheat the oven to 350 degrees.
1. Prepare four 8-inch cake pans: cut 4 rounds of parchment paper to fit the bottoms of the pans. Butter the pans liberally and position the parchment liners.
2. Sift together the flour, salt, allspice, nutmeg and cloves. Set aside. Push the blackberry and strawberry jams and peach preserves through a sieve into a bowl. Set aside.
3. In the large bowl of a heavy-duty mixer, cream the butter and sugar together until light and fluffy. Add the egg yolks and combine well.
4. Put the buttermilk in a 2-cup measure and stir in the baking soda.
5. With the mixer on low speed, add the flour and buttermilk mixtures alternately, beginning and ending with flour. Beat until batter is smooth. Stir in the jams, raisins and nuts.
6. Beat the egg whites until they form stiff, unwavering peaks. Scoop the whites over the batter and gently fold them in with a rubber spatula. Pour the batter into the baking pans, dividing it evenly and smoothing the tops with a spatula.
7. Bake in the middle of the oven 30 to 35 minutes, or until a toothpick inserted in the center comes out clean. Let cool 10 minutes on wire racks, then turn out on racks to cool completely. When completely cool, frost the cake with caramel icing.

CARAMEL ICING

Also known as seven-minute frosting, this icing dries out fast. Prepare it no longer than an hour before frosting the cake.

MAKES ENOUGH ICING FOR ONE LARGE CAKE

2 cups light brown sugar
1 cup granulated sugar
¼ teaspoon cream of tartar

¼ teaspoon salt
4 egg whites

1. In a pot or bowl set over simmering water, mix the sugars, cream of tartar, salt and egg whites. Over low heat beat steadily with an electric or rotary hand beater until the frosting stands in peaks, no more than 7 minutes.
2. Remove from the heat and continue beating until thick enough to spread.
3. To assemble, place one layer of the cake on a serving plate. Spread on about 1/2 inch of frosting. Top with another layer and spread; continue until the fourth layer. Spread the remaining icing over the top and sides of the cake. Drape the cake loosely with wax paper and set aside at room temperature for 1 to 2 days before serving.

Paul Cox with functioning whiskey still.

THE VARMINT ROAST

One of Marion County's exclusive "clubs" holds its meetings in a small country cabin on the outskirts of Lebanon. Up to thirty townsmen come together to "eat, talk politics, and tease," says Dick Moraja, "you know how a small town goes."

Like many honored institutions, the Varmint Roasts began informally, when Dr. Duncan Falot, the local eye, ear and nose physician, and others met on Wednesday evenings to cook whatever they'd bagged—rabbit, deer, elk, bear, coon, groundhog or fish. Some thirty people participate now, according to Dick Moraja. "Attorneys, farmers, guys working at G.E., retirees. You've got to work your way in. As people die, you move up the list."

These enlarged Varmint Roasts tend to be much tamer than in the past. "Sometimes we'll have ham, domestic rabbit or even chicken," says Dick, "cooked with turnips or kale."

HOW TO JUDGE A COUNTRY HAM

Finding a championship ham is a serious business, and it's not easy to fool an expert. "You can clean 'em up and dress 'em up to make them look better, but there's not much of a way you can affect meatiness and aroma," says Dr. James Kemp, a retired professor of meat science at the University of Kentucky and judge of the 1991 Country Ham Days Champion Ham Contest. "That was the best overall group I've seen here," opined Dr. Kemp, who judged the competition on conformation, workmanship, meatiness, color and aroma. (Taste is not a category, but a winner will surely taste wonderful.)

•*Conformation*: 20 points. The basic shape is pretty much a function of the pig. The hock can be long or short, but must be in proper proportion to plump meat.

•*Workmanship*: 10 points. The ham must be properly trimmed to "look nice," with just the right amount of pleasing fat and skin.

• *Meatiness*: 20 points. A good ratio of fat to lean is needed: too fat, the ham loses value; too lean, it can be dry and tough.

• *Color*: 20 points. Kemp prefers a uniform pecan brown.

• *Aroma*: 30 points. Determined by the nose, using experience guided by knowledge. An ice pick is inserted deep into the ham, then sniffed. The aroma must be "right," neither sour nor old.

Dick Moraja's ham, with an "excellent" 92 points, was top winner of fifteen hams. The ham fetched $600 at auction; the money went to the bidder's favorite charity.

SHELBY COUNTY FAIR AND HORSE SHOW

Shelbyville, Kentucky

FAIR DATE: FIRST FULL WEEK AFTER THE FOURTH OF JULY

GAZETTEER: *Shelbyville is between Louisville and Lexington. Fairgrounds on U.S. 60 west, off I-64. Downtown Shelbyville has become a hub of antique and collectible stores. This is horse farm country, for both race and American Saddlebred horses. Keeneland Racetrack and Churchill Downs, home of the Kentucky Derby, are nearby.*

"The person who knows most about the fair's history is Briggs Lawson," says Peggy Catlett, who has worked for the fair as executive secretary for some twenty years. "Briggs says he's been around so long he styles himself 'older than Absorbine Senior.'" "The horse show is an important part of our fair, and we're got a new covered arena for it, to replace the one that burned down," says Briggs Lawson. The fair was organized in 1842, and Mr. Lawson, who is active on the board, reckons that he has participated in a goodly number of them. "I recall the road horses, pulling buggies called roadsters, and how the crowd loved them," he says. "A fella named Bud James had two great horses, Bess B and Buster Tom. Everybody looked forward to Bud and those horses. My family had connections to American Saddlebred Horses–Shelby

Reverend James Edward Cayce and Judge Coleman Wright, manning the gates.

County calls itself 'Saddle Horse Capital of the World.' Those are the three- and five-gaited horses. The three-gaited do the walk, trot, and canter, and the five-gaited also do the slow gait and the rack. There's no thrill like the moment the judge calls out 'let your horses rack on!' and everybody screams and yells at that wonderful speed under saddle."

In addition to the agricultural events it shares with most fairs, the Shelby County Fair also has a big tobacco show. "This is the Burley Belt," says Jeanne Kemper of nearby Bagdad, who runs a dairy farm with her husband. "The golden leaf Burley gives cigarettes their flavor. It only grows here, nowhere else in the world. We raise nine and a half acres of Burley; our farm is one-third tobacco, two-thirds dairy. Without tobacco we'd be down and so would a lot of the other people around here. Milk prices are way low, and the dairy just carries us from week to week. At the end of the year, tobacco pays the bills. Without it, Shelbyville would be a ghost town.

"The average age of farmers here is over fifty. To start in the dairy business, you need a hundred thousand dollars. Most young people work their parents' farms—I don't know of any starting from scratch. We have factories taking good farmland. There's no resistance, 'cause everyone is willing to sell. When I was in school, it seemed we all had something to do with a farm, and we were comfortable with that. Now kids don't want others to know they live on a farm—they get teased. There's no pride in being from a farm anymore."

JEANNE KEMPER, PIE QUEEN

Despite the hard work and uncertainty, Jeanne loves the farm life. The family shows their Holsteins, "going to as many fairs as we can." Jeanne also loves to compete in baking contests, and has since she was a nine-year-old 4-H'er. She's a big-time winner, despite the occasional slip-up: "One year I baked fifteen pies to take to the State Fair, but I miscalculated and was a day late! I only realized my mistake at ten o'clock that night. We sure had a freezerful of pies."

KENTUCKY BOURBON PECAN PIE

Jeanne says, "The pecans and chocolate chips float to the top and the filling's real creamy."

MAKES 1 PIE

3 eggs
1 cup sugar
¼ cup melted butter
¾ cup light corn syrup
½ cup chocolate chips

½ cup chopped pecans
2 tablespoons Maker's Mark
     Kentucky bourbon
A 9-inch
     unbaked pie shell

• Preheat the oven to 350 degrees.
1. In a mixing bowl, beat the eggs and sugar. Gradually beat in the melted butter and corn syrup. Stir in the chocolate chips, nuts and bourbon.
2. Pour the mixture into the pie shell. Bake for 40 to 50 minutes until the filling is somewhat set.

JEANNE'S PIE CRUST

"I developed this crust for pecan and chocolate pies," says Jeanne Kemper. "The recipe makes enough for two pecan pies."

MAKES ENOUGH FOR A 9-INCH PIE OR TWO 9-INCH BOTTOM CRUSTS

3 cups all-purpose flour
1 tablespoon sugar
½ teaspoon salt
2 teaspoons cocoa powder
1 teaspoon ground cinnamon

¾ cup butter-flavored shortening
1 egg
½ cup milk
1 teaspoon white or cider vinegar

• In a large bowl, mix together the flour, sugar, salt, cocoa and cinnamon. Cut in the shortening, using your fingers or a pastry blender, until the mixture is crumbly. Stir in the egg. Slowly add the milk and vinegar, forming a large ball. Wrap in plastic and chill for at least an hour.

STRAWBERRY-RHUBARB PIE

"Almost foolproof," says Jeanne. This pie took the blue at the Kentucky State Fair in 1991.

MAKES 1 PIE

1½ cups sugar
3 tablespoons quick-cooking tapioca
¼ teaspoon salt
¼ teaspoon freshly grated nutmeg
3 cups (1 pound) rhubarb, cut into
    ½-inch pieces

1 teaspoon grated orange peel
1 cup sliced fresh strawberries
    Pastry for a 9-inch lattice-top pie
1 tablespoon butter, cut into bits

1. Combine the sugar, tapioca, salt and nutmeg in a large bowl. Mix in the rhubarb, orange peel, and strawberries, turning gently to coat the fruit. Let stand 20 minutes.
• Preheat the oven to 400 degrees.
2. Meanwhile, prepare the pastry for the lattice-top pie. Line the pie plate with the pastry, fill with the fruit mixture, and dot with the butter bits. Arrange the lattice top, and seal it. Bake for 35 to 40 minutes, until the filling is bubbling and the pastry golden brown.

APPALACHIAN FAIR

Gray, Tennessee

FAIR DATE: THIRD FULL WEEK IN AUGUST

GAZETTEER: *Gray is in Tennessee's northeast corner, 25 miles from the Virginia border and one hour's drive from Knoxville. Take I-81 coming from Knoxville; take exit 57A onto I-181, and take Suncrest Drive, exit 42 at Gray for the Fairgrounds.*

"Daniel Boone and Davy Crockett came from our part of the state," says Sherry Shadden, a fair staff member. "We're in the Great Smoky Mountains, and agriculture is number one with our fair. We also have a large Tennessee wildlife exhibit, with displays of native fish, snakes, birds and small animals in a natural habitat. We have an outstanding Farm and Home Building, and each year we present a theme to all our exhibits; one year it was 'Our Heritage, Land of the Cherokee,' and another year it was 'Melody of the South.' For the open judging of food, there are mirrors angled over displays so that the audience can see the food, and the judges talk about the items as they judge, giving out tips." Winners of the various divisions—food, horticulture, vegetables and needlework—are placed in the re-created old-fashioned town to the rear of the main stage.

"We take the young people's involvement in livestock showing very seriously," says Sherry. "In fact, we even have a category called 'Little Britches,' for children three to eight years old halter-leading calves six months old and younger. There are two divisions, the tiny kids, three, four and five, and the 'big' ones, six to eight. Their next stop is 4-H."

Sherry adds a bit of Tennessee trivia. "Ours is a very long state," she says. "Driving from here, you can be in Canada in less time than it takes to get to Memphis."

MINNIE PEARL FULKERSON

Mrs. Fulkerson lives on Fulkerson Road in Gray. Now in her eighties, she has worked over sixty years with the fair.

"My grandfather took up land at the head of Kendricks Creek in Washington County. He had a sawmill, and his lumber built the Harmony Baptist Church. My house is over a hundred and sixty years old. My husband, John Milton Fulkerson—he was a machinist and a jack of all trades—died in 1986 and I've lived alone ever since.

"When I was born, the census man happened by, and I didn't yet have a name. Mother was going to name me Peggy, but the two of them started talking. Mother

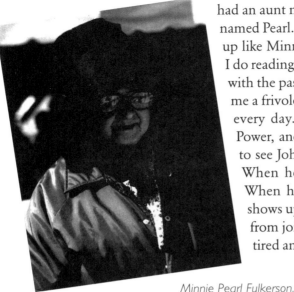

had an aunt named Minnie, and he had a daughter named Pearl. So that's my name! Nowadays I dress up like Minnie Pearl in the Grand Ole Opry, and I do readings. Here's one: 'I've become very social with the passing of the years. You might even call me a frivolous old gal; I'm seeing five gentlemen every day. As soon as I wake up there's Will Power, and he helps me out of bed. Then I go to see John. Then Charlie Horse comes along. When he's here he takes a lot of my time. When he leaves, there's old Arthur Itis. He shows up and spends the day, and he takes me from joint to joint. After such a long day, I'm tired and ready to go to bed—with Ben-Gay.'"

Minnie Pearl Fulkerson.

MINNIE PEARL'S ANYKIND FRUIT PIE

"The batter rises up through the fruit and makes a crust," says Minnie Pearl. "Blueberries make a wonderful pie."

MAKES 1 PIE

1 cup all-purpose flour	2 teaspoons vanilla extract
½ cup sugar	4 tablespoons (½ stick) butter, melted
1 teaspoon baking powder	2 cups lightly sweetened fresh fruit or
¼ teaspoon salt	canned fruit, drained
1 cup milk	

- Preheat the oven to 400 degrees.
1. Sift together the flour, sugar, baking powder and salt. Add the milk and vanilla and mix until smooth.
2. Melt the butter in an 8- or 9-inch pie plate. Pour on the batter but do not mix. Top with the fruit.
3. Bake in the center of the oven for 10 minutes. Lower the heat to 350 degrees. Bake about 50 minutes, until the pie bubbles and the crust is lightly browned.

FULKERSON TOMATO PUDDING

"Makes a nice side dish to any meal," says Minnie Pearl.

SERVES 4

2 cups canned tomatoes, drained and
   slightly broken up

1¼ cups well-crumbled toasted bread

¾ cup brown sugar

⅓ cup melted butter

1 teaspoon salt

½ teaspoon pepper

- Preheat the oven to 350 degrees.
- Grease a small baking dish. Mix the tomatoes, bread crumbs, brown sugar, butter, salt and pepper in a bowl. Transfer to the baking dish and bake until the pudding bubbles and browns, about 50 minutes. Let cool a little before serving.

WILSON COUNTY FAIR

Lebanon, Tennessee

FAIR DATE: MID-AUGUST, 8 DAYS

GAZETTEER: *The 106-acre fairgrounds are located at 945 Baddour Parkway, in Lebanon, 30 miles east of Nashville. Take I-40 coming from east or west. The town of Lebanon calls itself "Antiques Capital of the World." Most of the businesses on the town square are antique shops, and there are shops in nearby malls. Merchants strive to stock local Tennessee artifacts and antiques.*

"The newest feature at our wonderful fair is Fiddlers' Grove," says Karen Johnson, Ag Center secretary. "It's a reconstructed village of historical structures brought here from all over this part of Wilson County. The first blacksmith shop is here, and during the fair the Fiddlers' Grove Blacksmiths Association—people who are into old-time ironwork—fire up their forges and demonstrate making ornamental fixtures, bottle openers and other iron pieces. You can watch people churning butter, quilting, shearing sheep and making lye soap, too. We have a post office, a granary, a doctor's office, and an original town jail cell, with cast iron cots. The general store sells penny candy, hoop cheese and crackers, Cokes in the old 6½-ounce bottles, and Goo Goos, the marshmallow patty covered with caramel, peanuts and chocolate that's advertised on 'The Grand Ole Opry.' Recently we had campers for the fair from thirty-three states and Canada, and I can't tell you how many wanted to take Goo Goos home with them!

"One of the best things to come out of Fiddlers' Grove is the Fiddle Competition. We built a stage off the back porch of the general store. The first year, 1991, we had twenty-five entries. We even get stars, like John Hartford. If you know fiddle music, you know him. As our ads say, come to the Wilson County Fair, and celebrate good times."

WHAT'S A MEAL WITHOUT BISCUITS

"Being in charge of the baking contest is enlightening," says Jo Smith. "Lots of people don't know what the judges look for; we've even had people enter burnt pies! We wouldn't sample them because they didn't pass the appearance test, and those contestants were cross. Now I stress that food must look good."

Over six feet long! Over eleven feet around! Weighs approximately 3000 pounds.

Jo Smith has lived in Lebanon all her life. She grew up on a farm, lives on a farm, and has always worked with the fair: "My mother had me enter the 4-H competition, and almost every time I used that biscuit recipe, I won first place. When I grew up and married, one year I was champion of the quick bread division with that recipe. My children have also won 4-H competitions with these biscuits."

JO SMITH'S PRIZE-WINNING HOT BISCUITS

MAKES ABOUT 25 BISCUITS

¼ cup solid vegetable shortening

2 cups self-rising flour

⅞ cup less 2 tablespoons buttermilk

- Heat the oven to 450 degrees.
1. With the tips of your fingers or a pastry blender, cut the shortening into the flour until it resembles coarse meal. Add the buttermilk and stir in with a fork just until the dough holds together.
2. Turn the dough out onto a lightly floured work surface and knead a few times until smooth. Roll out 1/2 inch thick and cut out with a floured biscuit cutter. Transfer to baking sheets and bake 10 to 12 minutes, until golden.

MACARONI AND TOMATOES
"This is real good, real Southern," says Jo Smith.

SERVES 4

4 cups cooked macaroni

4 cups canned tomatoes, chopped

Salt and freshly ground pepper to taste

4 tablespoons (½ stick) butter,
   cut into bits

¼ pound Velveeta, cut into small cubes

½ cup milk

1 cup cracker crumbs

- Preheat the oven to 350 degrees.
1. Grease a 1¹/2-quart casserole. Put in the macaroni and enough tomatoes to fill two-thirds of the casserole. Season to taste with salt and pepper, add the butter, stir. Push the cubes of cheese down into the casserole. Add enough milk to top the dish. Do not stir! But poke a few holes to allow the milk to settle.
2. Top with cracker crumbs and bake until the crumbs are browned, about 40 minutes.

GEORGIA MOUNTAIN FAIR
Hiawassee, Georgia

FAIR DATE: BEGINNING FIRST WEDNESDAY OF AUGUST, FOR 12 DAYS

GAZETTEER: *A small town in northeast Georgia, Hiawassee is nestled in a lovely valley in the southern Appalachians. The town is about 100 miles north of Atlanta, 90 miles east of Chattanooga, on Route 76.*

This celebration of mountain crafts, heritage and bluegrass and gospel music is vitally important to the economy of little Towns County, whose leading industry is tourism. Despite its size, the fair is as straightforward and noncommercial as the proud mountain people who run it.

"What we've done is re-create a mountain village, to show the way of life here in the late 1800s and early 1900s," says manager Dale Thurman. "The houses are lit by kerosene lamps. We have a working blacksmith, cider made the old-timey way, and people cutting wooden shakes for roofing. You need a certain type of oak that grows here. They make a roof so that you can look up in a room and see the sky, but the roof won't leak. Amazing how that works. We have a little holler here with a sixty-year-old moonshine still, and people who know just how to make it. But we don't make it! Some years back the man who usually manned the still couldn't show up—seems this active moonshiner was spending a little time away, at the compliments of the government."

Trout smoked over charcoal with a little hickory and oak for flavor is a fair specialty; volunteer Jeff Waldroup smokes two hundred trout at a time, seasoned only with salt and pepper in the cavity. "I had smoked wild pigs at my getaway up in the mountains," he says. "Someone brought me a few trout, and at our board meeting, we came up with the idea of trout as something new. We also smoke ham and chickens. For the fair's twelve days we smoke sixty tons of food."

"Two-thirds of the fairgrounds is on property that my daddy once owned," says Joyce Holmes, who plants her garden by the zodiac. "Used to be small farms around here, but in 1941 the TVA built a dam just over the North Carolina line, and flooded the area. A lot moved away, but my daddy chose to stay here. He bought a spot and moved the house five hundred yards closer to the lake. He raised wheat and corn, and had a vegetable garden.

"I belong to the Garden Club, and we've put on a flower

Cooking hominy.

show since 1951, when the fair started. I'm a retired teacher, and we, the Towns County Retired Teachers, host the mountain home and school in Pioneer Village.

"Planting by the zodiac signs is what I do. A book called *The Ladies' Birthday Almanac* tells you how. If you make kraut when the moon is dark, your kraut is white; any other time, it will darken. I make my kraut in a crock. I use firm cabbage heads, salt to taste, put on clean cabbage leaves, and weight them down with white rocks. I tie a clean cloth over, put it in a dark place and let it sour for about three days. Then I take it out and squeeze out the brine, pack it in a can and cook it twenty minutes to seal. I do it in the dark of the moon, and my kraut stays white.

"Back in the old days, people strung green beans on a string and hung them up to dry; it was a form of dehydration. When they wanted beans in the wintertime, they'd break 'em off, soak 'em overnight, and cook 'em with a piece of meat, just like fresh beans. These are called 'leather britches.'

"People used their wits to survive. When the moss grows heavier on the north side of the tree, when the hornets build closer to the ground, when the corn has a thicker shuck, they knew there'd be a bad winter, and plan for it."

SENATOR RICHARD B. RUSSELL'S SWEET POTATO PUDDING

"We have a lot of company," says Joyce. *"This dish was a favorite of Georgia's long-time senator."*

SERVES 6, OR 12 AS PART OF A BUFFET

3 cups mashed sweet potatoes	1 tablespoon vanilla extract
1 cup granulated sugar	1 cup light brown sugar
2 eggs, beaten	⅓ cup all-purpose flour
8 tablespoons (1 stick) butter, at room temperature	8 tablespoons (1 stick) butter, melted
	1 cup chopped pecans

• Preheat the oven to 350 degrees.

1. Butter a casserole dish. Mix the sweet potatoes, granulated sugar, eggs, the room-temperature butter and vanilla together and put in the dish, smoothing the top.

2. Make the topping. Combine the brown sugar and flour, mix in the melted butter, and spread over the potato mixture. Sprinkle on the chopped nuts. Bake about 40 minutes, or until the top is browned.

LAKE COUNTY FAIR

Eustis, Florida

FAIR DATE: EARLY APRIL, FOR 10 DAYS

GAZETTEER: *Lake County is the area called the Golden Triangle, 30 miles north of Orlando. The county has over 1,500 named lakes. Once a seat of the citrus industry, Lake County is now largely recreational and home to retirees.*

"A fair's not like a sale, where if you don't sell it today, you can put it back in inventory," says Betty Dittman, veteran fair manager who has run the Lake County event since 1985. "So much depends on the weather. We have only ten days to make a profit, and then the fair is history."

Betty, who is president of the Florida Federation of Fairs, has made her career in a business that demands tough-mindedness and split-second decisions. "When I was managing the upper Carolina State Fair," she recalls, "a showman had a python with a stomach ache. Seems it needed a shot of penicillin. I can tell you, it was a job persuading the local vet to deal with a fifteen-foot snake!"

The Lake County Fair is over seventy years old, having grown from a small school fair to its present eminence. Livestock exhibits are still prominent, though the county's agricultural base was almost destroyed in the Big Freeze of '87. Grove after grove of frozen, blighted citrus trees were plowed under. "Before the freeze, I'd be driving home from Daytona, and the blossoms smelled so beautiful," Betty mournfully recalls. Happily, beef production, produce and renewed citrus groves are now thriving.

People at both ends of the age span—senior citizens, of whom there are many in the area, and kids—get a lot of attention at the Lake County Fair. "We have the Kitchen Band," says Betty. "Seniors with bazookas and improvised instruments, like the base made out of a washtub and rubber bands. And for the children, we have an exceptional petting zoo." There's King Arthur, a llama, and the ducklings, Prince Charles and Lady Di. (Are they still together? Don't ask.)

Betty's been in the fair business some thirty years, and she knows the headaches of bringing together hundreds of 4-H exhibitors, carnival people, entertainers, rides, games and expectant visitors. "The fair may have made me gray on top," she says with a laugh, "but it's kept me young at heart. Lord, may I live till I'm a hundred, so I can keep seeing all the good the fair does."

GAIL NORRIS' EGGPLANT CASSEROLE

Gail's late husband Charles was the fair president, "off and on for some twenty years." Their son Happy has also served as president. "My happiest times are in my kitchen," Gail says. She recommends serving this eggplant casserole with a boneless rolled pork shoulder, coated with coarse salt and black pepper and cooked for about four hours over an indirect charcoal fire. "Baste the roast every 20 minutes with equal parts of white vinegar, Worcestershire and soy sauce. It could be done in the oven but would not have the same good flavor."

SERVES 4 TO 6

1 large eggplant, about 2½ pounds	A long dash Worcestershire sauce
1 large onion, finely chopped	A good dash Tabasco
3 tablespoons butter	½ cup grated mild cheddar cheese
1 teaspoon dried thyme leaves, crumbled	2 cups broken saltine crackers
1 egg, lightly beaten	Salt and freshly ground pepper to taste

1. Peel the eggplant and cut it into 1-inch cubes. Cook the cubes, partially covered, over moderate heat in unsalted water to cover, about 25 minutes. Drain.
• Preheat the oven to 350 degrees.
2. Meanwhile, cook the onion gently in the butter with the thyme until softened and translucent but not brown, about 8 minutes.
3. Combine the eggplant, onion, egg, Worcestershire sauce, Tabasco, cheese, cracker crumbs, and salt and pepper. Transfer to a greased casserole and bake uncovered 50 to 60 minutes, or until the casserole is bubbling.

Mounted bass exhibit, 1937.

A FAIR MANAGER'S PRAYER FOR WEATHER

Please God, keep the skies free of rain, crowds at the gates,
days free of trouble, and money left over.
—Betty Dittman
Lake County Fair

CITRUS COUNTY FAIR

Inverness, Florida

FAIR DATE: THIRD WEEK OF MARCH FOR 6 DAYS

GAZETTEER: *The area has seven rivers, with a wildlife park and a manatee sanctuary. Good fresh- and saltwater fishing. The fairgrounds are located at 3600 S. Florida Avenue, Inverness, 70 miles north of Tampa on Highway 41.*

Best of show, long ago.

"Like everywhere else in central Florida, the Big Freeze in the early eighties knocked out most of our citrus," says fair manager Jean Grant. "On Christmas morning I ran out to pick a grapefruit, and it froze to my hand. Some farmers are going to strawberries, now, and this is a big watermelon area. According to our Florida fair laws, an agricultural fair like ours must produce a certain number of agricultural events, and we are big in steers, heifers, swine and goats. We have more animals than produce, as our fair is in March, and that's not prime growing time. We focus on FFA and 4-H, and we promote family unity. No alcohol, but we do have a carnival.

"Most of the time the fair runs smoothly, but I guess everybody has to have one day that's the worst in your life. I sure remember mine. Before the gates opened, we had a Dumpster full of elephant dung to get rid of. Too late, I realized I'd sent the driver the wrong way, and he crashed through the top of a septic tank. There was the truck, in this tank like a cement coffin! I had to get a crane to lift it out. What an odor!

"We opened to the public with a steer show. First thing, a steer got loose and threw a young man—I sent for an ambulance, but before it arrived we were having our famous toilet toss, folks seeing how far they could throw chamber pots. One slivered, and a young man cut his leg. So I sent for a second ambulance. Then a real tragedy happened—a young father who'd taken antihistamines died on the Tilt-a-Whirl ride; I had to find his wife and children and soothe the child who'd been on the ride with him, and that meant a third ambulance.

"I turned around, and someone had put something in our huge fish tank, and all the fish were belly up. Just then an irate wet lady came in, crying that she'd been on the Ferris wheel, and 'water came out of noplace,' and drenched her. Turned out a vendor who called himself Roger Rabbit had a stand where people shot at water balloons and it was placed too near the Ferris wheel. Speaking of rabbits, a little boy stuck his finger in a cage of exhibition rabbits and got bitten; I had to call the University of Florida to inquire if the rabbit could be rabid—the answer was no. Just as I got the boy and his father calmed down, in came the rabbit's owner, asking

'Where's the idiot who put his finger in the cage?' I hope I never have a day like that again."

PEGGY HARTMAN'S SOUR ORANGE PIE

"Most of the oranges in Florida are grown on trees that have been grafted onto hardy sour orange root stock," says Peggy Hartman, secretary of the Citrus County Extension Service. "If the buds freeze, the tree reverts and produces only sour oranges. They're real tangy—but only good for cooking. If you can't find sour oranges, use a combination of regular oranges and lemons." Peggy won ribbons with her pie at the Citrus County Fair, In 1986 and 1978. Here it is, with slight adjustments.

MAKES 1 PIE

1 cup granulated sugar	1½ cups fresh sour orange juice, strained,
5 tablespoons cornstarch	or 1 cup orange juice
½ teaspoon salt	and ½ cup lemon juice
1 cup water	4 egg yolks, beaten
2 teaspoons grated sour orange peel,	2 tablespoons unsalted butter
or 1 teaspoon each grated lemon	A 9-inch baked pastry shell
and orange peel	

THE MERINGUE:

4 egg whites	½ cup superfine sugar
1 teaspoon sour orange or lemon juice	

1. Whisk together half the granulated sugar, the cornstarch and the salt. Add the water, grated peel and juice, and whisk until smooth. In a saucepan, bring the mixture to the boil over moderate heat, stirring constantly until it begins to thicken. Remove from the heat.

2. With an electric mixer, beat together the remaining 1/2 cup granulated sugar and the eggs yolks until light in color. Beat in 1/2 cup of the hot sugar-juice mixture, and add all of the yolk mixture to the contents of the saucepan. Stir over moderate heat until thick. Take care that the mixture does not boil. Remove from heat, stir in the butter, and pour the filling into the pie shell. Let cool.

• Preheat the oven to 350 degrees.

3. Meanwhile, make the meringue topping. Beat the egg whites and the teaspoon of juice to soft peaks. Gradually beat in the superfine sugar, beating until stiff peaks form and all the sugar has dissolved. (Test with your fingers.) Spread the meringue over the filling, touching the edge of the pie shell. This will prevent shrinkage.

4. Bake 12 to 15 minutes, until the meringue is golden brown. Cool thoroughly before serving.

HERE COMES THE CARNIVAL!: STRATES SHOWS

"My grandfather came from Greece as a boy," says Sibyl Strates, marketing director of the Strates Shows. "He became a world-class wrestler, and during the off-season, he traveled with carnivals in what were called 'athletic shows.' He was known as 'Young Strangler Lewis,' and for five dollars he'd wrestle all comers. In 1923 he and two partners organized their own carnival business. Shortly after the lion, Old King, bit off the giraffe's head, he found himself going it alone. In 1929 he bought railroad cars to carry the show up and down the Eastern Seaboard."

Today the Strates Shows travel eight months a year, in fifty-eight railroad cars and twenty-six trucks. At winter quarters in Orlando, the railroad siding is a mile long. "We carry three to five hundred people at a given time," says Sibyl, "and we have some fifty rides, including a hundred-and-twenty-six-foot-high Ferris wheel. Setting up the show and breaking it down are fine-tuned operations. We can start to break at midnight, and at 8 a.m. the next day we're ready to go.

"My parents met when my father, James, who runs the company, was in the Marine Corps. My mother was a Baptist from Indiana, and found herself on the road, surrounded by passionate Greeks! She adapted to the close quarters of living in a private railroad car, and to Greek cooking, with lots and lots of people dropping in for meals.

"Living on a train is great fun. As kids, we all went along when school was out. We're five children, and four of us work in the carnival. Sleeping on the train is wonderful; it's as comfortable travel as you could want. Today my brother Jim and sister Susan travel the season with their families, but they prefer to live in roomy, custom-fitted vehicles.

"The animals, freak shows and girlie shows are a thing of the past; today people want high-tech, high-speed rides, like the Orbiter and the Zyclon roller coaster. People still like bumper cars; we just got new ones, and they fold up neatly into one trailer. We also have a double-decker carousel, made in Venice. I do remember when we had musical revues, which were like what Las Vegas shows are now. We once had a

All aboard! Mr. and Mrs. James E. Strates.

beautiful all-girl troupe from Japan; it took them an hour to get into their elaborate kimonos. As a little girl, I'd go and watch them get ready."

PHYLLIS STRATES' DOLMATHES WITH AVGOLEMONO SAUCE (STUFFED GRAPE LEAVES WITH EGG AND LEMON SAUCE)

"Stuffed grape leaves are our holiday dish," says Phyllis, Sibyl's mother. "I have five kids, so I make a lot of them. We have them several times a year, and always at Christmas and Easter."

MAKES 60 TO 70 DOLMATHES

1 (16-ounce) or 2 (8-ounce) jars grape leaves	1 large egg
¾ cup raw white rice	2 tablespoons salt, or to taste
1½ pounds chopped sirloin	1 teaspoon freshly ground pepper, or to taste
2 finely chopped onions	Avgolemono Sauce (p. 120)
1 tablespoon dried mint leaves, crumbled	

1. Trim the coarse veins and stems from the grape leaves. Rinse them well and drain them.
2. Mix together the rest of the ingredients except for the avgolemono sauce. Cover the bottom of a heavy pot with a layer of small or torn grape leaves.
3. Place a teaspoon of filling in the center of each leaf and fold like an envelope. First, fold over the bottom, then the sides. Then roll from bottom to top. Be sure to fold lightly, as the filling will swell as it cooks.
4. Arrange the stuffed grape leaves tightly, side by side, in the pot, seam side down. Make as many layers as needed. Cover the top layer with another layer of coarse leaves.
5. Cover the dolmathes with cold water and put a plate on them to keep them firmly packed down. Place over high heat and bring to the boil. Immediately lower the heat, partially cover with a lid, and let the dolmathes simmer for 40 to 45 minutes, or until tender when pierced with a skewer. (Taste one to make certain it is cooked to your liking.) Serve hot with avgolemono sauce. The dolmathes are also good cold.

AVGOLEMONO SAUCE

3 large eggs, at room temperature ½ teaspoon salt
  Juice of 2 lemons   Dash of white pepper

1. Beat the eggs with a whisk until frothy. Beat in the lemon juice, salt and pepper. Gradually beat in 1/2 cup of the warm, but not boiling, liquid the dolmathes have cooked in. Return the mixture to the pot with the dolmathes. The liquid in the pot should not be too hot, or the egg-lemon sauce will curdle.

GREEK BROWN BUTTER SPAGHETTI

"My mother-in-law taught me this quick and delicious dish," says Phyllis Strates.

SERVES 4

8 ounces spaghetti   Salt and freshly ground black pepper
8 tablespoons (I stick) butter   to taste
    ½ to I cup freshly grated Parmesan cheese

1. In a large pot of salted water, boil the spaghetti according to package directions. Drain, and place on a large serving platter.
2. As the spaghetti boils, cook the butter over low heat in a small skillet. Watch it carefully as it begins to turn brown; do not let it darken too much or it will turn bitter. Immediately pour the butter over the spaghetti and toss to coat. Add the salt and pepper and 1/2 cup of the cheese. Pass the rest of the cheese at the table.

DADE COUNTY YOUTH FAIR & EXPOSITION

Miami, Florida

FAIR DATE: MID-MARCH, FOR 18 DAYS

GAZETTEER: *The address of the sixty-acre fairground is 10901 Coral Way (SW 24th Street), Miami.*

"Since 1952, our fair has been a showcase for youthful achievement. We now have some fifty thousand displays, from cooking to cattle raising to photography," says Nell Ohff, director of student exhibits and a board member of the Florida 4-H Foundation. "We're involved with FFA and 4-H, which are lead-ins to good citizenship. Our beginnings were agricultural, and much of our focus still is—South Florida has a great deal of agriculture—but we want urban kids, too. Students from five to twenty-one are evaluated by professionals in their areas of interest; they don't compete against each other. Everyone gets a prize, and many get scholarship money."

It's provident that the fairgrounds are blacktopped, with big buildings and shower facilities; perfect for the three thousand National Guard troops that occupied the site during the disaster of Hurricane Andrew, which devastated much of Dade County on August 24, 1992. "We were proud to have them," says Nell. "They were wonderful people and helped us all.

"The storm brought new meaning to words like 'patience,' 'electricity,' and 'neighbor,'" she says. "I spoke to so many people in South Dade County, and found a new and profound attitude concerning priorities and togetherness. People yelled out of what had been a window or door, 'Come on over and help us eat the meat in the freezer before it spoils.' The last remark made by one of my colleagues to a friend before the storm hit was 'Save the cow!' They had purchased the 1992 Dade County Youth Fair grand champion steer, which was in the friend's freezer. With the coming of Andrew, and the going of electricity, we found neighbors barbecuing in large numbers, and doing it together. Many new and special recipes emerged from the storm, along with many new friendships."

PORK CHOPS HURRICANE ANDREW

"I'm giving this recipe in the spirit in which it was created," says Nell Ohff. "It gives you the sense of how brave people cope in a disaster." The chops should taste just as good under less harrowing circumstances.

SERVES 6

About two cups of peeled and chopped pineapple or apples, grapes, or whatever is in the refrigerator

½ cup or more sugar

Salt and pepper

6 medium pork chops

Worcestershire sauce

1. Build a good fire and let it burn until the heat is moderate. Set a grill over the fire.
2. While the fire burns down, mix the fruit and sugar and let stand until liquid is released, about 30 minutes. Form a sturdy pan from aluminum foil, heavy-duty if possible. Lightly salt and pepper the chops and place them on the foil. (Be sure you have used clean water to wash the chops; 2 drops of household bleach per gallon of water will do.) Sprinkle the chops with drops of Worcestershire sauce and pile the fruit over the chops. Use more foil to completely seal the packet. Place on the grill and cook for 1½ hours, without opening.

After Hurricane Andrew;
National Guard on the Fairgrounds.

"WE WILL REBUILD" THANKSGIVING DINNER

*This meal was put together by South Dade County, Florida, survivors. Hurricane Andrew
hit on August 24, 1992. By Thanksgiving, the area was still a disaster area,
in everything but human spirit.*

1 to 10 small turkeys	Milk
Sweet potatoes	Apples
Butter	Brown sugar
Cinnamon	Raisins
Pancake syrup	Approximately 12 neighborhood
Baking potatoes	children
Stuffing mix	Lots of foil and foil trays
Chopped onion	3 or more gas or charcoal grills
	Large bunch of neighbors

- Prepare in the following order:
1. Place the turkey(s) on spits and set over moderate grill fire. Assign children (or moms, dads and grandparents) 15-minute shifts to rotate turkeys. Thirty minutes from each person should do it.
2. Cut sweet potatoes in half, skin on, and place in foil trays. Add butter and cinnamon, cover with syrup. Double wrap with foil and place over low fire on second grill.
3. Rub butter on potatoes, wrap in foil and place on another grill.
4. Mix stuffing mix with chopped onion and milk until moist. Place in foil trays, wrap in foil, and place on second or third grill.
5. When done, send out a shrill call for everyone to eat! A different prayer is given on a Thanksgiving Day like this one: Gratitude for new friends and camaraderie, and a general thanks that only material goods were lost. We still have the important ingredients that make up a good life.
6. After dinner, when the coals have settled down, core the apples and fill the holes with a mixture of brown sugar, butter, raisins and cinnamon. Place upright in foil pans, wrap with foil. Place on the coals and gently cook until tender.
7. There are no television parades or football games, so-o-o-o-o, gather around the fire. Pick up sticks for toasting marshmallows. Colanders wrapped in foil make great corn poppers, but be sure to *wire* on a large stick for the handle; friction tape burns. Baked apples, popcorn and marshmallows. Maybe some neighbors still have a little brandy left. *Life is good!*

ROBINSON'S RACING PIGS

Racing pigs are not new—a number of fairs have featured them—but it took Carlota and Paul Robinson to take hog hurtling into the big time. Back in 1984 two fair officials mentioned to Paul, a fair entrepreneur, that they'd seen pig racing at the Montgomery County Fair in Gaithersburg, Maryland (page 82). "Cuter than sin," they said. "Heck," said Paul, "Carlota and I can do that." "It started as a joke," adds Carlota, "and it changed our lives."

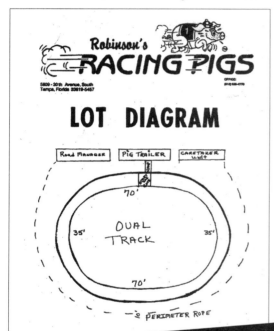

The Robinsons, who are based in Tampa, now have ten units on the road. Each one consists of a horse trailer converted to a traveling porcine palace ("These pigs live a life of luxury," says Carlota); a track; a ramp from the trailer to the track; and flags, banners and signs. And the pigs. To a musical fanfare before the race, the swinemaster of ceremonies picks cheerleaders from the audience, one to an entry. The pigs wear numbered racing silks (actually, place mats held on with double-faced carpet tape).

Kentucky Derby hopefuls may Run for the Roses, but Robinson's pigs dash for the Oreo. ("Pigs have insatiable sweet teeth," says Carlota.)

See 'em fly!

Hog Heaven. Secretariat at six months. (She now weighs 340 pounds.)

Four little pigs at a time scorch around a 150-foot oval track at a thundering fifteen miles per hour; the race takes about seven seconds. The winner gets a whole cookie, and the others get crumbs. The race goes by so fast that a second race—a steeplechase over hurdles this time—is run immediately. The crowd goes wild! "Come on, Hammy Faye Bacon!" "Faster, Arnold Schwarzenhogger!" Then it's all over, and the winner is Roseanne Boar Arnold, by a snout.

"Pigs are outgoing animals, and they love the spotlight," says Carlota. "They smile to the audience. Whoever coined that expression 'ham' for actors must have spent a lot of time around pigs."

CARLOTA ROBINSON'S WEIRD CHILI

*"You know how you run out of things and make do? Well, that's what I did.
I invented this dish when we were on the road with the pigs. The cocoa makes it 'weird,' but
don't omit it. It adds a certain something."*

SERVES 8

1½ pounds ground chuck	1 tablespoon ground cumin
1½ pounds lean pork sausage, removed from casings	1 tablespoon paprika
	Freshly ground pepper to taste
2 large onions, chopped	2 (15-ounce) cans kidney beans
6 cloves garlic, minced	2 (15-ounce) cans tomato sauce
2 tablespoons sugar	2 (15-ounce) cans crushed tomatoes
1 tablespoon unsweetened cocoa powder	1 small can tomato paste, diluted with
¼ cup chili powder (hot! If you desire, use less)	1 can of water
	1 (12-ounce) can beer

1. In a large stewpot, cook the meat and onions over moderate heat until the meat loses its red color. Drain off the fat and return the pot to the stove.
2. Stir in the remaining ingredients. Cover the pot and simmer 4 hours, or all day long. Check often and lower the heat if the chili boils; if the liquid is cooking away too fast, add a little water. Serve with garlic cheese toast.

NATIONAL PEANUT FESTIVAL AND FAIR

Dothan, Alabama

FAIR DATE: USUALLY THE FIRST WEEK IN NOVEMBER, FOR 8 DAYS, CLOSED ON SUNDAYS

GAZETTEER: *Dothan is in the southeast corner of Houston County, eleven miles from both Florida and Georgia. This tri-state area is commonly called the Wiregrass, named for local vegetation. Coming from the south, take 231 North, and exit at Ross Clark Circle. The fairground is at Cottonwood Road and Ross Clark Circle.*

"This used to be cotton country," says Carrie Cavender, administrative assistant for the National Peanut Festival Association, "but the boll weevil was so destructive that farmers switched from cotton to peanuts. That turned out to be a blessing in disguise. One little town in the Wiregrass called Enterprise put up a monument to the boll weevil, for doing the area a favor!"

The festival began in 1938, and fittingly honored the man who devised myriad uses for the peanut: Dr. George Washington Carver of Tuskegee Institute. Dr. Carver's memory was honored in 1993, the fiftieth anniversary of his death. The festival is big, and so is the Wiregrass area's peanut crop—some three hundred thousand tons per year. "Our main purpose," says Carrie, "is to celebrate the peanut farmer." Just in case anyone forgets, baskets of peanuts are everywhere; roasted, fried, raw and boiled. Roasted peanuts are free. The Peanut Festival is a complete agricultural show with judging in all livestock categories, and includes a calf scramble for 4-H and FFA youths ages twelve to fifteen. Winners who catch calves take them home, halter-break them and bring them back to show at the following year's festival. "That event always gives us all a warm feeling," Carrie says.

The week before the fair, Miss National Peanut Festival Queen and Little Miss National Peanut Festival are selected. The royal ladies work throughout the year to represent the industry. "I was only seventeen when I was elected queen in 1974," says Debra Boyd. "I was a small town girl from Enterprise, who bagged groceries and pumped gas. I'd never been on an airplane. Suddenly I was in Scottsdale, Arizona, watching Governor Carter announcing his bid for the presidency, and I found myself dining with Senators in Washington, representing Alabama as well as peanuts." Debra's daughter Katie was the seven-year-old Little Miss Peanut in 1989, making them the first Queen and Little Miss in one family. "The Peanut Festival wants a natural little girl, not a dolled-up one," says Debra. "I cut ribbons for the

fair and the big parade,"
adds Katie. "I also greased
the pigs for the Pig Scramble
with peanut oil, to make
them hard to catch." Katie,
who has enjoyed a year of
celebrity, has always known
what she wants to be when
she grows up—a doctor.

*Katie Boyd, Little Miss National
Peanut, and her mother Debra,
former Miss National Peanut
Festival Queen.*

ANNE DAWSEY'S OLD-TIMEY PEANUT BRITTLE

*"I was raised on a peanut farm," says Anne Dawsey, 1991 Volunteer of the Year.
"When I was a little girl we saved the seed peanuts from one year to the next. We'd shell
them in winter and plant them in spring. Because of import-export laws, every farm has its
allotment from the government for how many peanuts you can plant. If you want to plant
more, you must buy somebody else's quota." Anne has a high regard for the peanut, and
has entered the recipe contest every year. She's had a number of winners.*

SERVES 6

1½ cups sugar	2 cups shelled raw peanuts
¾ cup dark corn syrup	¼ teaspoon baking soda
½ teaspoon salt	

1. Butter a large baking sheet.
2. In a heavy saucepan combine the sugar, corn syrup, salt and peanuts. Cook over moderately high heat until the peanuts crackle and pop. Stir in the soda.
3. Immediately spread the mixture on the baking sheet. When cool, break into pieces. Store in an airtight container.

FANTASY CHEESECAKE

Martha Jones of Dothan was the 1991 grand prize winner. "I'd been entering the contest for about five years, and took honorable mentions and awards like that. But it was my fantasy to win the big prize. And I did!" Martha, a nursing student, is married and has three sons.

MAKES 1 CHEESECAKE

THE CRUST:

1¾ cups graham cracker crumbs

⅓ cup melted butter

¼ cup granulated sugar

½ cup chopped honey-roasted peanuts

THE CAKE:

3 (8-ounce) packages cream cheese, softened

2 teaspoons vanilla extract

3 eggs

1 cup sugar

1 cup sour cream

12 ounces chocolate morsels

1 cup chopped peanuts

3 ounces semisweet dark baking chocolate

½ package caramel candies (about ¾ cup)

⅓ cup water

• Preheat the oven to 350 degrees.

1. Make the Crust: Combine all the crust ingredients in a bowl. Press the mixture onto the bottom and side of a 9-inch springform pan. Set aside.

2. To make the cake: Beat the cheese, vanilla and sugar until the mixture is creamy. Beat in the eggs, one at a time. Fold in the sour cream. Fold in half of the chocolate morsels. Pour into the springform pan and bake 60 to 70 minutes, until the center is set. Turn the oven off, leave the oven door ajar and let the cake cool for 1 hour. Refrigerate the cake for at least 4 hours.

3. Remove the side from the pan. Top the cake with the remaining chocolate morsels and peanuts.

4. Over low heat, melt the caramels and baking chocolate in the water. When smooth, dribble the mixture over the cake with a spoon.

NESHOBA COUNTY FAIR

Philadelphia, Mississippi

FAIR DATE: FIRST FULL WEEK OF AUGUST, EXCEPT EVERY 4 YEARS BEFORE STATE PRIMARIES; THE FAIR IS THEN HELD DURING THE WEEK BEFORE THE FIRST PRIMARY

GAZETTEER: *Neshoba County, in mideastern Mississippi, is two hundred miles from Memphis. Philadelphia, the county seat, is the only town in the county, which contains a large Choctaw reservation. The fairgrounds are located on Highway 21, nine miles southwest of Philadelphia.*

Neshoba County Fair has horse racing, a rousing midway, agricultural displays and all the many attractions of a thriving rural fair. It has more, too. Founders' Square, which regulars claim is the "finest sittin' room in the South," is a serious stump for politicians; then-Governor Reagan kicked off his presidential campaign at Neshoba. But what really sets Neshoba, a spot "in the middle of nowhere" apart is the small, jammed-to-the-rafters city on the grounds that flourishes for only one week in the year.

"Mississippi's Giant Houseparty" is perhaps an odd nickname for a fair, but then no other fair is like Neshoba's. Officially organized in 1891 as an educational agricultural meeting, the fair drew people from considerable distances in this sparsely settled county. Gradually, it seemed a good idea to come and stay for the fair week. People began to build cabins. Fast friendships grew. Now, year after year, folks come not only from Philadelphia, but from all over the South, and even from Europe. Today the more than six hundred one- and two-story cabins are part of each family's heritage; every cabin crammed to the porch swing with as many as forty happy campers all eating together and sleeping in relays; it's a place where adults "visit" as only Southerners know how, where children forge the bonds that will bring them back year after year, and where they intend to bring their own children to live the short summer idyll.

"When my son was in the army," recalls Mary Grace Johnson, the doyenne of the Neshoba Fair, "he got leave to come home from England for the fair week. He'd rather come home for the fair than for Christmas, he said. My husband, Norman, figures that more than eight thousand people live here in fair week. These cabins aren't vacation houses, they're just for the fair. I started coming as a bride in 1946. I had a two-burner kerosene stove, and not enough electricity for refrigeration. Now we have central plumbing, air conditioning upstairs in our cabin, and a freezer. Norman was president of the fair for twenty-six years, and we really do like company. Lunch is the big meal, and it goes on for several hours.

"We have three meals a day at our cabin, and we've had as many as a hundred and fifty people to lunch during fair week, so I have to plan, even though I never know how many we'll have. I am sure that most of us spend more for the fair than we do for Christmas. We have plenty on the table, but never wasted food. It's a solid week of talking and walking, without TV, just visiting and being together. Everything is planned around the Neshoba County Fair. We think from fair to fair, almost."

Mrs. Norman Johnson, Jr., marks centennial.

FRIED CATFISH

"Catfish farms are all over down here. Traditionally, we serve this with coleslaw and hush puppies," says Mary Grace Johnson.

SERVES 4 OR 5

¾ cup yellow cornmeal	2 teaspoons salt
¼ cup flour	4 or 5 catfish fillets
Dash of ground red pepper	Vegetable oil for deep frying
¼ teaspoon garlic powder	

1. Combine the cornmeal, flour, ground red pepper, garlic powder and salt. Spread it out on wax paper. Dip each fillet in the mixture, shaking off excess.
2. Pour about 3 inches of vegetable oil into a heavy 12-inch skillet. Heat, slowly, to 350 degrees as registered on a deep-frying thermometer. Add the fillets, two at a time, and fry until golden brown, turning once, about 6 minutes for thin fillets. Drain on paper towels.

HAMBURGER NOODLE CASSEROLE

"Casseroles like this one are useful for serving the many guests we have at the fair,"
says Mrs. Johnson.

SERVES 8

1½ pounds fairly lean ground beef	½ cup sour cream
1 tablespoon butter	⅓ cup chopped onion
2 (8-ounce) cans tomato sauce	1 to 2 teaspoons chopped green pepper
1 teaspoon salt	¼ teaspoon ground red pepper
1½ cups cottage cheese	8 ounces wide noodles, cooked
8 ounces cream cheese	

• Preheat the oven to 350 degrees.
1. Brown the ground beef in the butter. Stir in the tomato sauce and salt and remove from the heat.
2. Combine the cheeses, sour cream, onion, green pepper and ground red pepper. Butter a 3-quart casserole and spread out half the noodles on the bottom. Add the cheese mixture, then the remaining noodles. Layer on the ground beef mixture. Bake for 30 minutes, or until bubbling.

MARY GRACE JOHNSON'S NESHOBA FAIR PRUNE CAKE

"We start cooking and freezing for a month before the fair," says Mrs. Johnson.
"My prune cake is one I've taken to Neshoba for many a year."

MAKES 1 CAKE

1 ½ cups vegetable oil
1 ½ cups sugar
3 eggs
1 teaspoon baking soda
1 cup buttermilk
2 cups all-purpose flour

1 teaspoon freshly grated nutmeg
1 teaspoon allspice
1 teaspoon salt
1 teaspoon vanilla extract
1 cup chopped cooked unsweetened prunes
1 cup chopped pecans

THE ICING:

1 cup granulated sugar
½ cup buttermilk
½ teaspoon baking soda

8 tablespoons (1 stick) butter
1 tablespoon dark corn syrup
1 teaspoon vanilla extract

• Preheat the oven to 300 degrees.

1. Cream the vegetable oil and sugar together until fluffy. Beat in the eggs, one at a time. Stir the soda into the buttermilk and set aside.

2. Sift together the flour, nutmeg, allspice and salt. Beat lightly into the egg mixture. Add the vanilla. Then add the prunes, pecans and the buttermilk mixture. Combine, pour into 2 buttered and floured 9-inch pans, and bake for 35 minutes, or until a toothpick inserted in the center of the cake comes out clean. Cool the layers on racks.

3. Make the icing. Cook all of the ingredients together over a moderate heat until thickened, about 20 minutes. (To check, drop a small amount into cold water; it should form a soft ball.) Ice the cake while icing is still warm.

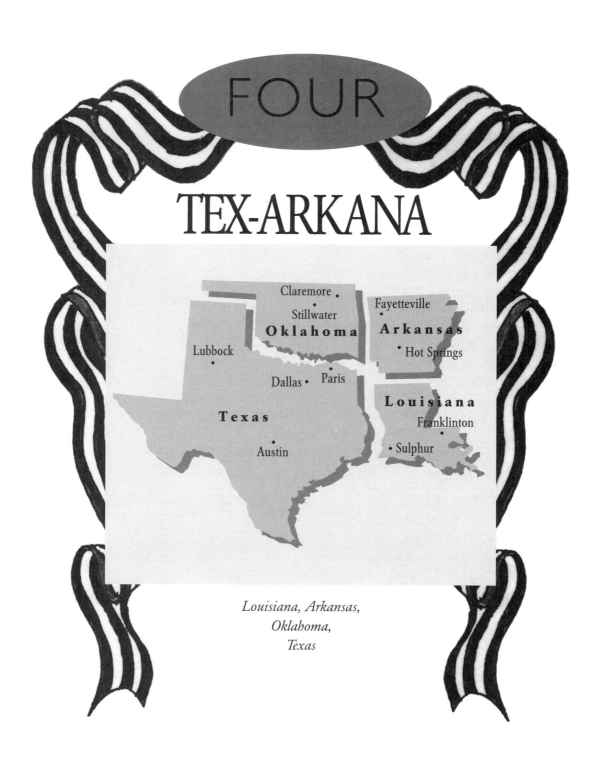

FOUR

TEX-ARKANA

Claremore

Stillwater

Fayetteville

Oklahoma

Arkansas

Hot Springs

Lubbock

Dallas • Paris

Louisiana

Texas

Franklinton

Austin

Sulphur

Louisiana, Arkansas,
Oklahoma,
Texas

CALCASIEU-CAMERON FAIR

Sulphur, Louisiana

FAIR DATE: FIRST MONDAY AFTER LABOR DAY, FOR 10 DAYS

GAZETTEER: *The fair serves two parishes, Calcasieu and Cameron. Fairgrounds are located just north of Sulphur on LA. 27, the DeQuincy Highway. This is Cajun country, on the Gulf of Mexico.*

Queen Bercy LaFleur, 1935.

The Cal-Cam Fair, as it is known, owes its origins to a medical man's good will. Dr. A. H. LaFargue, on his house calls, obliged his patients by carrying produce from one to another. Finding that his chores as unofficial agricultural middleman were getting in the way of his practice, in 1922 "Doc" LaFargue sparked the idea of a fair where farmers and the public could meet, letting him get on with the job of curing the ill and delivering babies. (In forty years of practice he figured that he had delivered about five thousand of them.)

Doc LaFargue would be pleased that one of the fair's oldest and most popular features is concerned with health. It's the Better Baby Contest; nurses judge the contestants, and winners must be absolutely in the pink. "There's nothing cute about this contest at all," says Sheila Rather, a fair official, "but the judges do look for personable babies that smile and glow. It's so humid and sticky here in Louisiana that you're proud if you can keep a baby healthy."

As at most fairs, queens are important. The fair decided to honor the first queen, sixty years after her reign, by making her Grand Marshal of the 1992 fair.

Pageant, Open Air Auditorium, early 1950s.

First queen Bercy LaFleur was delighted to crown the incoming queen. Back in 1932, the rival contestants had to collect a penny for each vote, the money to go to the Fair Association. Miss Bercy had the edge, "But my daddy was so afraid I wouldn't win he gave me ten dollars. I ended up with two hundred and eighty-six dollars, way ahead of my rival, partly because every driller at the sulfur mines outside of town gave me fifty cents. In those days fifty cents would buy two pounds of round steak. That amount of money was amazing during the Great Depression."

Miss Bercy well remembered the king she had selected: "His name was Precious Hughes. I never knew his first name, but the nickname fit." Unfortunately, Precious was too infirm to attend Miss Bercy's gala reunion.

Being Cajun country, the Cal-Cam food booths offer fare not seen elsewhere. "We have catfish courtbouillon, fried alligator meatballs, and boudin—stuffed with what we call dirty rice, which is made with chicken livers—in a sausage casing. It looks funny, but oh! it's good!" says Sheila.

CATFISH COURTBOUILLON

Fair manager Wilridge Doucet is "a real Cajun; I didn't speak English till I went to school. I learned to cook watching my mother." Retired now, Wilridge coached the football team and taught social studies for twenty-seven and a half years. He likes hunting, fishing and cooking Cajun on special occasions.

SERVES 4 TO 6

2 tablespoons vegetable oil	1 (16-ounce) can tomato sauce
2 pounds catfish fillets	1 cup chopped scallions (green onions)
1 cup chopped onion	1 tablespoon Tabasco, or to taste
10 large cloves garlic, peeled and chopped	Salt and freshly ground black pepper to taste

1. In a large skillet, heat the oil over moderate heat. When it sizzles put in the catfish. Add the chopped onion, garlic and tomato sauce to the pan. Lower the heat, cover the pan and simmer gently for 10 minutes
2. Stir in the scallions and Tabasco, and add salt and pepper to taste. Simmer, covered, 20 minutes longer. Serve over rice.

PORK AND SAUSAGE JAMBALAYA

Note the amount of Tabasco. "We like it hot!" Wilridge says. Non-Cajuns should use discretion.

SERVES 4 TO 6

1 tablespoon vegetable oil	½ cup chopped scallions (green onions)
3 cups diced lean pork	2½ cups water
1 pound pure pork sausage, cut into	1 cup long-grain rice, rinsed
1-inch rounds	1 teaspoon salt
½ cup chopped onion	1 teaspoon Tabasco, or to taste
2 cloves chopped garlic	

1. In a large cast-iron Dutch oven or heavy skillet, heat the oil over moderate heat. Cook the pork and the sausage, turning occasionally, until brown. Pour out most of the fat. Add the onion, garlic and scallions and cook, stirring from time to time, until the onion is translucent, about 6 minutes.

2. Add 1/2 cup of the water, raise the heat and bring it to the boil, while scraping up the browned bits from the pan. Add the rice, salt, Tabasco and the remaining 2 cups of water. Bring to a second boil. Lower the heat, cover, and cook over low heat for about 35 to 40 minutes, until the rice is cooked and most of the liquid absorbed.

Queen, waiting for parade to begin, 1950s.

WASHINGTON PARISH FREE FAIR

Franklinton, Louisiana

FAIR DATE: THIRD WEDNESDAY IN OCTOBER, 4 DAYS

GAZETTEER: *Washington Parish is in the "toe of the boot" of Louisiana, in the southeastern part of the state. (In Louisiana, the word "parish" is used instead of "county.") Coming from the south, take Highway 25; from east and west, take Highway 10. The 45-acre fairground has many tall pines and spreading oak trees.*

"Lots of volunteers give of their time, talents, blood, sweat, and tears to make ours one of the most exceptional county fairs in the whole nation," says super-fair volunteer Virginia Killingsworth. "We're very proud that we charge no admission. The entertainment on our outdoor stage is free. You pay only for the rodeo, the midway rides and food."

The fair began in 1911, part of the effort to educate farmers in ways of diversifying crops; cotton, a notorious soil-depleter, was on its way out. The first fair was a raging success (so many people came that some of the livestock had to be tied in side streets) and not without its lighter side: "One of the most interesting events of the day was pulled off by a young lad, who brought in a common cow with which he entertained the crowd by making the cow walk on her knees, forward and backward, lie down, stand on her hind feet, and other things," according to a report in the local newspaper.

Mile Branch Settlement, a pioneer village complete down to privy and smokehouse, is named for the creek that runs through the fairground. All the structures, excepting the replica Half Moon Baptist Church, were rescued from various parts of the country, brought to the site, and restored. People can book the Big Kitchen for family reunions: "As many as one hundred, sometimes," says Mrs. Killingsworth. "We have a lot of big clans, like the McGees. When I came here as a bride in 1941, the first six women I met were all Mrs. McGee! My head was spinning.

"During the 1920s, Mr. Martin K. Penton and Miss Roxie E. Alford got married on the stage at the fair. The wedding was very elaborate, and the purpose was to show the country people how a formal wedding was held. A few years ago we re-created that wedding, and the person who performed the ceremony was the Pentons' son."

Virginia Killingsworth and Marco, Indian Encampment, Mile Branch Settlement.

LUVERITY DAVIS' CHICKEN PIE

"This dish is a specialty of Franklinton, and recipes can differ," says Virginia Killingsworth. "Luverity Davis has been making the pie for me for years; she's one of the best cooks anywhere. You can make it in the roasting pan, or you can divide it among aluminum foil bread pans, and freeze until needed."

Note: The "dumplins" are actually strips of pie dough.

SERVES 8 TO 10

1 (3- to 3½-pound) chicken, cut into pieces, or equivalent parts	4 cups all-purpose flour
Water to cover	1 teaspoon salt
8 tablespoons (1 stick) butter	1⅓ cups vegetable shortening
1½ teaspoons salt	About ¾ cup very cold water
½ teaspoon freshly ground black pepper	2 cups rich milk or half-and-half

- Preheat the oven to 350 degrees.
1. Place the chicken pieces in a 12 × 9 × 3-inch roasting pan or flameproof casserole. Cover with water, add the butter, salt and pepper. Cook the chicken, partially covered, over moderately low heat until tender, about 40 minutes.
2. Meanwhile, make the dumplins. Mix the flour and salt. Cut in the shortening with a pastry cutter until the mixture resembles coarse meal. Slowly add the water, a little at a time, using only enough to hold the pastry together. Form into a ball and refrigerate for 30 minutes.
3. Roll out enough pastry to cover the top of the roaster or casserole. Roll out the rest and cut it into strips, about 3 inches by 2 inches. These are the dumplins.
4. Remove the chicken pieces from the broth and set aside. (At this point, you can remove the skin and debone the chicken pieces, if you like.) Bring the broth to a rolling boil and add the dumplins one at a time so that they do not stick together. Return the chicken to the broth and stir gently until the broth begins to thicken. Add the milk, and stir. If the liquid has not thickened enough, cook over high heat for a few minutes to reduce it. Top with the crust and pinch it over the sides of the casserole.
5. Bake the pie in the center of the oven for 45 minutes, or until the crust is golden. Serve with fluffy white rice.

WASHINGTON COUNTY FAIR

Fayetteville, Arkansas

FAIR DATE: SATURDAY BEFORE LABOR DAY, FOR 5 DAYS

The county is in the heart of the Ozark Mountains, in Arkansas' northwest corner. The many clear lakes and fresh mountain air draw tourists and retirees. In fall, the foliage resembles that of New England. The fairgrounds are two miles from the University of Arkansas, home of the famed Razorback football and basketball teams. For the fairgrounds, take Highway 71 bypass.

"I came here as a student, married a rancher, and I've been on the fair board for some twenty years," says Doris Cassidy, the Washington County Fair president. "We try to keep a homey touch," she says. "For good sports who give it a try, there's an 'I Milked a Cow' badge. Our six food booths are all run by community groups, not concessions. The 4-H clubs make a mean taco salad, the Goshen community serves up fried chicken dinners, and the Grange does terrific beans and corn bread.

"We put other young people to work, too. The FFA kids serve on traffic control, parking and cleanup. They have a good time as they earn money for their projects, and learn skills, too. As one of the agricultural instructors told me, 'It certainly beats selling candy door to door.'"

"We're a complete fair from an agricultural point of view, but as we are in the center of chicken-raising country, our poultry exhibit is really large. Poultry has its own building."

ROOT! ROOT! RAZORBACKS!

During late August and early September, over the great state of Arkansas may be heard fevered yells of "Whoo Pig Sooiee!" Out-of-staters and other foreigners take this as the call of the mountain man collecting his hogs, but as every modern-day Arkansan knows, this is the call to arms of the mighty Razorback football team, "Get 'em, Hogs! Whoo Pig Sooiee!!"

ARKANSAS BREAKFAST CASSEROLE

Born and raised in Washington County, Kathryn Skelton has been active with the fair for over thirty years and is chairman of the Women's Division. "The fair's the first date I mark on my calendar," she says. As for her hearty casserole, "It's good for a company brunch."

SERVES 6

6 slices bacon, cut into 2-inch pieces
1 small green pepper, cored, seeded and chopped
2 tablespoons finely chopped onion

6 cups cubed cooked potatoes
½ cup shredded sharp cheddar cheese
6 eggs, lightly beaten
Salt and pepper to taste

1. In a large heavy skillet, cook the bacon over moderate heat until crisp. Remove the bacon to drain. Pour off all but 3 tablespoons of dripping.
2. Heat the dripping remaining in the pan till it sizzles. Add the pepper, onion and potatoes and cook, stirring gently, until the potato cubes are browned.
3. Sprinkle on the cheese and stir until it melts. Add the eggs and cook, stirring gently, until the eggs are set. Season with salt and pepper. Sprinkle on the crumbled bacon and serve at once.

GARLAND COUNTY FAIR

Hot Springs, Arkansas

FAIR DATE: FIRST MONDAY AFTER LABOR DAY, FOR 6 DAYS

GAZETTEER: *The fairgrounds are located just off Martin Luther King, Jr., Expressway, Route 88, in the historic spa town of Hot Springs. It is a mountainous area of great natural beauty. There is beef and poultry raising, little other agriculture.*

"The Garland County Fair is rich in boisterous history," says A. J. Parker, who has been on the fair board for many years. "A group purchased the present grounds in 1941, forty-seven acres of scrub oak and sour grass. Volunteer labor did the job. I was in high school in 1945, and I remember the original beef cattle barn was put up by the Lakeside FFA chapter, and veteran ag students. They used scrap lumber. Did a good job, too.

"Back at the end of the Forties, the fair was in bad shape. Some crooked politicians got involved. Rented tents, then sold them! Raffled a car—the winner got the keys, but no car! Since then, no politicians have been involved.

"That old Roger Miller song, about 'ridin' down a dusty road, going to the county fair,' takes me back. In the Forties and Fifties, for us the greatest thing was going to the fair in the pickup truck. I recall time spent getting the animals ready, sleeping on the ground between cows—cows will not step on you. When it got cold, you'd scooch up to a cow to keep warm.

"Did I ever see President Clinton at the fair? Can't say I did. But he grew up in Hot Springs, and he played the saxophone in the high school band, so he must have come to the fair like other boys.

"The fair has grown tremendously; huge home economics exhibits, a demolition derby and a wonderful carnival, the same carnival for twenty-five years. Our carnival people are of the upper crust. And a fair isn't a fair without a rodeo. But if we didn't have the swine, the sheep and poultry and rabbits—we'd lose the heart."

"It's been a tradition since the Forties to have Cooperative Extension Home-makers' Concessions here at the fair," says Susan C. Neeper, county Extension Agent and home economics leader, who has worked with the fair for twenty-three years. "We make the food—like our chili—from scratch, and we're famous for it. President Clinton attended the fair many times. During one of his gubernatorial campaigns, he made hamburgers in our old kitchen.

"I guess I've loved fairs since I was in the first grade, and my teacher entered a picture of a dog I had drawn. I won a blue ribbon! It meant a lot to me then, and it still does. We do the judging before the fair opens, and a few years back a young homemaker came in to check on her jar of mixed vegetables. She couldn't find it, and she was beside herself. Well, we located that jar in the sweepstakes section, which is the best of hundreds of entries. When all of the superintendents and workers saw the surprise and happiness on her face when she realized her first entry ever had won—why, we were really, really glad. That's the kind of thrill that lasts."

Demolition Derby.

THE DEMOLITION DERBY

The all-American pastime of grown men in stripped-down junkers whacking into each other in a demented real-life game of Dodgems is an enormous crowd puller at many a county fair. "It's family oriented," says Kenneth Orrell, who promotes the Demo Derby at the Garland County Fair in Hot Springs, Arkansas. "They come to see blood. That's what they want to see. But we haven't had anybody (knock wood) get hurt.

"The object," says Orrell, "is to get out there and disable a car by whatever means you can, other than knocking in the driver's door. You can hit him front, right, knock off his wheels and fix him so he can't start. He has two minutes to get his car running, then he's disqualified.

"We hold four heats a night for three nights of the fair, eighteen cars per heat. Nine cars line up on one side of the rodeo arena, and nine cars on the other, all facing out. The crowd gives a five-count countdown, and when the announcer hollers 'Go!' they back out, and the race is on. It's a free-for-all. This is about my nineteenth year promoting the Derby, and that's as close as I want to get."

James Mattingly, who is on the Garland County Fair Board, is a competitor. "My strategy is to hit head on in front, and back up and hit if they're behind me. I've been banged a lot—busted a helmet like a watermelon on a door jamb. Figured my neck'd be sore, but it wasn't. With a hit on the side you don't take much shock, but an unexpected hit in the back can pop you around pretty good. A good ram is one hundred feet; you can get up to forty-five m.p.h.

"You try to build a good car so it will last. There's nothing on the inside but the driver's seat. Everything is down to the bare metal; no chrome, fiberglass or glass. You can't add to the car, you can only take away. There is no reinforcement, no welding. You have a five-gallon gas tank, and if a heat lasts as long as forty-five minutes, you can run out of gas.

"On Saturday night there's an extra heat, and all heat winners participate in that— twelve in all. Each heat pays a hundred dollars, and the big prize for the last Derby heat is five hundred dollars, so the winner of that comes out with six hundred dollars. But I'm not in it for the money. They call me Smiley, 'cause I laugh a lot."

EXTENSION HOMEMAKERS DOUGHNUT MIX

These doughnuts without holes are not too large; each one makes a satisfying little snack.

MAKES 3 DOZEN

3¾ cups all-purpose flour	¼ cup powdered milk
1½ teaspoons salt	2 eggs
½ teaspoon ground ginger	1 cup water
1½ teaspoons grated nutmeg	¼ cup vegetable oil, plus additional for
1 tablespoon ground cinnamon	deep frying
5 teaspoons baking powder	Confectioners' sugar (optional)
1 cup granulated sugar	

1. Sift together the flour, salt, spices, baking powder, granulated sugar, and powdered milk. Beat together the eggs, water and 1/4 cup of the oil. Beat the wet ingredients into the dry ingredients.
2. Fill a heavy wide pan with 3 inches of vegetable oil. Heat slowly to 375 degrees on a deep-fry thermometer. Drop rounded teaspoons of batter into the hot fat. Fry on one side until the batter bubbles; turn once. Fry for approximately 6 minutes. Drain the doughnuts on paper towels. Sprinkle with confectioners' sugar, if desired.

EXTENSION HOMEMAKERS' COOPERATIVE EASY CHILI FOR A CROWD

"This recipe was developed by us to serve at the Extension Homemakers' Concession at the Garland County Fair. The longer it cooks, the better it tastes!" says Susan Neeper.

SERVES 10 TO 12

2 large onions, chopped	1 (8-ounce) can tomato sauce
2 tablespoons vegetable oil	2 cups water
3 pounds ground beef	Salt and pepper to taste
3 (1-pound) cans kidney beans	¼ cup chili powder
3 (1-pound) cans tomatoes	2 tablespoons cumin

1. In a large heavy pot, cook the onions in the sizzling oil until limp. Add the meat and brown, turning often and stirring to break up the lumps. Drain off the fat.
2. Meanwhile, puree the beans and tomatoes in batches in a food processor. Add to the meat mixture, along with the tomato sauce and water. Season with salt and pepper, and stir in the chile powder and cumin. Bring to the boil, lower the heat, and let the chili simmer, partially covered, for at least 1 hour.

ROGERS COUNTY FREE FAIR

Claremore, Oklahoma

FAIR DATE: SECOND WEEKEND AFTER LABOR DAY, FOR 4 DAYS

GAZETTEER: *Rogers County is named for Clem Rogers, father of humorist Will Rogers, who claimed Claremore as his home town. It is the site of the Will Rogers Memorial: the Will Rogers birthplace is twelve miles north of town. Other attractions are the J. M. Davis Gun Museum and the Lynn Riggs Memorial, honoring the author of* Green Grow the Lilacs, *a novel that became the musical* Oklahoma! *The fairgrounds are off Highway 20.*

"Rogers County is a place of rolling plains with lots of prairie, and much horse production," says fair board member Rick Reimer. "Horse racing is a big part of the fair at Will Rogers Downs, here at the fairgrounds.

"Food is important! Our Extension Homemakers feed hundreds from their kitchen; 4-H kids for breakfast, who've spent the night in the barn with their animals, and all the superintendents and judges. The eleven Homemaker groups in the county each make four dozen cinnamon rolls and eight pies. Much of the food is made in advance, like the noodles for the Sunday chicken and noodle dinner. They're frozen till needed."

Will Rogers and Blue Boy, State Fair, 1933.

"I was born in Cuba and grew up in Key West" says Esther Stipp, "and we've lived in Italy and Spain. My husband was transferred often, once to Claremore. Of all the places we've been, we decided to retire here. The people are so caring, so nice. I love to cook, and I joined the extension homemakers. In Key West there are no cows; they imported one once a year so the children could see what a cow looked like. And here I am in cowboy country!"

WILL ROGERS:
"I NEVER MET A HAWG I DIDN'T LIKE."

Will Rogers (1879–1935), trick roper, vaudeville and screen star, syndicated columnist and wry commentator on the foibles of his countrymen, was perhaps the leading celebrity of his day. Joseph H. Carter, director of the Will Rogers Memorial in Claremore, Oklahoma, relates Rogers' fair activities:

"A Hampshire boar named Blue Boy shared stardom with Will Rogers in his 1933 movie classic, State Fair. *When it was suggested that he purchase Blue Boy after the movie was completed, Will Rogers said, 'I wouldn't feel right eating a fellow actor.'*

"Part of the plot of State Fair *concerns a food competition. The wife of Rogers' character has secretly added brandy to her mincemeat; not knowing what she's done, Rogers has also given the mincemeat a heavy second shot. Needless to say, the delicious mincemeat wins the blue ribbon!*

"About food, Will Rogers wrote in his column of September 25, 1932: 'I love my navy beans better than any other dish or half-dozen dishes. Just plain white navies, cooked in plenty of ham or fat meat with plenty of soup among 'em. Not catsup or any of that other stuff. Just beans and cornbread, old corn pone (white, with no eggs) with the salt and water it's cooked with, and raw onion. Those three things are all I want.'"

ESTHER STIPP'S YELLOW SQUASH SUMMER SOUP

Esther says to chill the bowls the soup is served in.

SERVES 8

4 tablespoons (½ stick) butter
2 tablespoons vegetable oil
1 cup finely chopped onion
2 cloves garlic, finely chopped
3 pounds yellow summer squash
    (crookneck), thinly sliced

4 cups chicken stock
1 cup half-and-half
1½ teaspoons salt
½ teaspoon white pepper
    Finely chopped parsley, for garnish
8 thin slices squash, for garnish

1. In a large heavy pot, over moderate heat, heat the butter and oil. When sizzling, cook the onion and garlic, stirring often, until they are soft and translucent. Add the squash and chicken stock, and simmer until the squash is very tender.

2. Transfer the soup in batches to a blender or food processor. Add some of the half-and-half to each batch. Blend until very smooth. The soup should be fairly thick. Place in a bowl and stir in the salt and pepper. Serve chilled, garnished with a little parsley and a slice of squash.

CRISPY CHICKEN-SHRIMP ROLLS

"Here is our favorite company dish," says Esther Stipp. The dish is a delicious spin on the better-known Chicken Kiev.

MAKE 8 SERVINGS

8 boneless skinless chicken breast halves	2½ teaspoons salt
1½ cups finely chopped shrimp	1 teaspoon fresh lemon juice
12 tablespoons (1½ sticks) butter,	Vegetable oil for frying
at room temperature	1 cup all-purpose flour
¼ cup finely chopped scallions	1¼ teaspoons baking powder
(green onions)	¾ cup water

1. Carefully pound each chicken breast to a thickness of about 1/4 inch; avoid making holes. (Place each breast between two large sheets of wax paper and pound with a wooden mallet or heavy bottle.) Refrigerate until needed.

2. In a bowl, blend together the shrimp, butter, scallions, 1 1/2 teaspoons of the salt, and the lemon juice. Form the mixture into a rectangle, about 8 × 3 × 2 inches. Wrap in plastic wrap and chill until firm.

3. Cut the shrimp butter into 8 finger-size pieces. Lay a pounded chicken breast in front of you, wide side toward you. Lay a piece of shrimp butter on the wide end; roll up, jelly roll fashion, tucking in the sides as you go. (I use a little dab of flour to glue the edges together if they get stubborn.) Secure them with toothpicks if you like. You should have 8 neat cylinders. Chill them for 20 minutes or so, to firm them up.

4. Pour 1 inch of oil into a wide heavy pan. Over moderate heat, bring the oil to 370 degrees on a deep-fry thermometer. As the oil heats, whisk together the flour, the remaining teaspoon of salt, the baking powder and the water.

5. Using tongs, dip 4 of the chicken rolls into the batter to coat, letting excess drip off. Lower them into the hot oil. Fry until golden, turning occasionally, about 12 to 15 minutes. Drain on a baking sheet lined with paper towels, placed in a 200 degree oven. Repeat with the 4 remaining rolls. Remove the toothpicks, if you have used them, and serve at once.

PAYNE COUNTY FAIR

Stillwater, Oklahoma

FAIR DATE: LAST PART OF AUGUST, EARLY SEPTEMBER, FOR 5 DAYS

GAZETTEER: *Stillwater is an agricultural center, home of Oklahoma State University. It is equidistant from Tulsa and Oklahoma City. The fairgrounds are three miles east of Stillwater, off Highway 51.*

"I was elected to the Payne County Fair board in 1935, and I served for fifty-eight years, when I stepped down as chairman," says D. H. Fisher. "The fair is free, and anyone is eligible to participate. We have an extra-nice facility on what used to be the county poor farm, where indigent old people lived. Been some time since we've had that."

"An event at our fair is called 'Power of the Past,'" says Mary Silvers, a fair board member and longtime supervisor of open classes in the women's division. "It features demonstrations of old threshing and washing machines, can crushers, rock crushers, soap making, all the crafts of the past. We formerly had horse pulling, but that ran its course. Now we have lawn and garden tractor pulls. You take a plain old riding mower and soup it up. There's very precise placing of weights, and lots of competition. Now these are grown men and ladies! But it's a lot of fun. There's a tractor driving contest too; both written and field tests. They have to back a tractor and wagon, which is very difficult. Some very good drivers find they can't back. People from all over the state come for the cattle penning. The contestants, three at a time, have to cut the animals out of a herd of ninety and bring them to a pen; they have three minutes to do it. We also have old fiddlers and gospel music, and we have lots of good local talent."

Ladies' Exhibit Building, 1935. Dresses and hooked rugs.

MARY SILVERS' POTATO SALAD

"This salad is best after a day or two," says Mary. "That gives the flavors a good chance to blend."

SERVES 8 TO 10

THE BOILED DRESSING:

4   eggs, well beaten
½ to 1 cup sugar
4   teaspoons flour
1   teaspoon salt
½   cup water
½   cup cider vinegar
1   tablespoon butter
1   teaspoon prepared mustard

THE SALAD:

10   medium potatoes, boiled, cooled
       and peeled
4   hard-boiled eggs
½   cup chopped onion
3   good-sized sweet pickles,
       finely chopped
     Small jar diced pimiento
1   teaspoon celery salt, or more,
       according to taste

1. Make the dressing: Place the eggs, sugar, flour, salt, water, vinegar, butter and mustard in a saucepan and mix well. Bring the mixture to the boil, lower the heat and simmer, stirring constantly, until the dressing has thickened. Cool slightly.

2. Make the salad: Cube the potatoes and dice the hard-boiled eggs. Combine in a bowl with the onion, pickles, pimiento and celery salt. Pour the dressing over, and mix well. Cover with plastic wrap and chill overnight or longer.

RED RIVER VALLEY FAIR

Paris, Texas

FAIR DATE: WEEK BEFORE LABOR DAY, 6 DAYS (OPEN EVENINGS ONLY, ALL DAY SATURDAY)

GAZETTEER: *Paris, in Lamar County, is 110 miles north of Dallas on the Oklahoma border, near Idabel. The Women's Christian Temperance Union was organized in Paris in 1882. From Dallas, take I-30 North.*

"Our fair began in 1911. In 1916, a disastrous fire destroyed most of Paris, and some people took refuge on the fairgrounds," says Rita Jane Haynes, president of the fair and also 1992 president of the Texas association of Fairs and Expositions. "The Fair Association did everything they could to help the citizens of a town in ashes."

Paris' citizens take major upheavals in stride. "Back in 1982," Rita Jane recalls, "we were having a fundraiser for our new Community Exhibit Center. Our fair committee arranged to have a well-known local person 'kidnapped' by the police and put in a 'jail' we'd arranged, until people donated enough money to set him free.

Patriotic trick riders.

"Well, the kidnappee was from the radio station. He said, 'Let me out, there's a tornado coming!' 'You're just saying that to get out!' the kidnappers scoffed. That went on awhile. Then someone ran out from the radio station and yelled, 'Hit the road! There really is a tornado coming!' So once again, the fair people stepped in. They stopped taking money and started helping folks out of their homes."

The Red River Fair features a rodeo: trick riders who plant a foot atop the backs of *two* galloping horses; and a barnyard full of baby animals. A daytime event is held for preschool children, who can pet the animals, see a puppet show, and squeal at a clown's antics.

The only way you can get the Paris natives down, it seems, is to jerk them in the mud, at the losing end of a tug-of-war rope. Local businesses happily field five-member teams. Winners hoist a trophy; losers wallow in the mire.

RICK POSTON'S CHILI

Rick is a director of the Red River Valley Fair Association, and a card-carrying chilihead. Chiliheads are special people who will go wherever necessary to compete against fellow zealots. Rick competes in twenty to twenty-five chili cookoffs each year. In 1992 this recipe took seventh place among two-hundred and fifty-four entries at the Chili Appreciation Society International Cookoff in Terlingua, Texas, the championship of the chili world. Both the ingredient and instructions lists are quite specific. Rick requests that you cook in "dumps" (see below).

MAKES JUST OVER 2 QUARTS (36 OUNCES)

1 tablespoon Crisco
2½ pounds chuck tender, cut into
   3/8 x 3/8-inch cubes

• In a large heavy Dutch oven or kettle, heat the Crisco until sizzling. Brown the meat cubes, stirring often. Pour off accumulated juices, and add the first DUMP. Cook the chili covered through all stages and DUMPS, and stir it frequently. (If additional liquid is needed during cooking, add more beef broth.)

DUMP 1:

2 (14½-ounce) cans Swanson's
    Beef Broth
1 teaspoon beef bouillon granules
1 teaspoon chicken bouillon granules
1 teaspoon Pendery's El Rey Chili
    Pepper (see Note)

4 teaspoons Pendery's High Color
    Paprika
5¾ teaspoons onion powder
½ teaspoon garlic powder
1 (8-ounce) can tomato sauce
3 serrano chili pods, pierced

• Cook for 1 hour. Squeeze the chili pod juices into the pot. Discard the pods.

DUMP 2:

½ teaspoon Pendery's 40 K
    Cayenne Pepper
½ teaspoon Sazon Goya
¼ teaspoon ground white pepper
¼ teaspoon ground black pepper

2 teaspoons ground cumin
1 teaspoon garlic powder
1 tablespoon Pendery's Original
    Chili Powder
2 tablespoons Gebhart's Chili Powder

• Cook 1 hour 25 minutes, then add:

DUMP 3:

½ teaspoon Sazon Goya
½ teaspoon onion powder
1 tablespoon Pendery's Original
    Chili Powder

2 teaspoons ground cumin
⅜ teaspoon Pendery's 60 K
    Cayenne Pepper
2 tablespoons Gebhart's Chili Powder

• 5 to 10 minutes before turn-in (presentation to judge or eager guests), add:

KICKER:

⅜ teaspoon salt
1½ teaspoons ground cumin
1½ teaspoons Gebhart's Chili Powder

• Stir well, dish up. Very, very rich!

NOTE: Pendery's does an extensive mail order business. Write Pendery's Inc., at
1221 Manufacturing, Dallas, Texas 75207-6505, or call 1-800-533-1870.

KATHLEEN WILLIAMS' BAKED POTATO SALAD

Kathleen, a home economics teacher at North Lamar High School, is Creative Arts Co-Chairperson at the Red River Valley Fair; she's been associated with the fair for twenty-five years. This recipe was a District 4-H Food Show winner.

SERVES 6

4　medium potatoes, peeled and boiled until barely tender, diced

½　pound Velveeta cheese, cubed

½　pound sharp cheddar cheese, cubed

½　cup chopped onion

Salt to taste

1　cup Miracle Whip salad dressing

½　pound bacon

1　cup stuffed olives

1½　cups plain croutons

1. Butter a 9 × 13 × 2-inch glass baking dish. Layer the ingredients: first the potatoes, then the cheeses, then the onion. Sprinkle with salt and cover with the salad dressing. Cover the dish with plastic wrap and chill overnight.
2. Next day, cook the bacon until crisp. Crumble it and scatter over the dish. Arrange the olives over the casserole and sprinkle on the croutons. Bake, covered, for 25 minutes. Uncover and bake 5 minutes longer, or until slightly browned.

• Preheat the oven to 350 degrees.

STATE FAIR OF TEXAS
Dallas, Texas

FAIR DATE: LATE SEPTEMBER OR EARLY OCTOBER, FOR 24 DAYS

GAZETTEER: *Historic Fair Park is on Route 30, minutes from downtown Dallas.*

Whooo-eee! This is the big one, all right! But what else would you expect of Texas? The fair sprawls over two hundred and seventy-seven acres, has some two hundred and seventy-six buildings, the world's largest Ferris wheel, a nightly electrified twilight parade with floats, bands and drill teams, a marine marching band, and an attendance of over three million each year. Add to all that the Cotton Bowl, scene of the annual fierce pigskin rivalry between Texas and the Sooners of Oklahoma. Over all towers the fifty-two-foot figure of Big Tex, wearing a ninety-gallon hat and size seventy boots. His clothes are stitched by tentmakers. Big Tex's thundering "Howdy, folks!" is the signature of the fair.

"I think of the Creative Arts Building as an oasis in a mostly commercial area," says Elizabeth Peabody, its director. "We have twenty-seven food contests during the fair, and over fifty judges. Anyone who can cook stands a chance of getting a ribbon."

"I was a top competitor for many years," says Lillie Crowley, "both in food and in sewing. God has blessed me with many talents. Now I am privileged to receive the entries in the canned goods division, and to display the prize winners. I also do a country cooking demonstration, using what's at hand.

Tallest Ferris wheel in the western hemisphere.

The Cotton Bowl; Texas battles Oklahoma during the Fair, before 72,000 fans.

"All four of my grandparents emigrated from what is now Czechoslovakia, where they were citizens of Austria. They came in through Galveston, and went to farming in Ennis County. My husband, Lee—his name's Paul, but we call him Lee—is a retired long-distance truck driver. When the children were grown, I traveled with him on our rig, and loved every minute of it. I got to see things in this beautiful country of ours that I didn't dream possible: the dogwoods in the Carolinas, the redbuds in Tennessee, and the Louisiana swamps you can see deep into. Going north, we'd follow the berries in the fields. I can spot wild poke from here to Washington, D.C. Poke is the spring green that God gave us before the cultivated greens. It is considered a body cleanser, and it will work you out good; everything that ails you. I cook it like spinach and throw off the first water, then scramble it into eggs. Poke must be cooked, and it can't be cultivated; it must pass through a bird to take root.

"We are organic farmers on two acres, and we take our surplus to the farmers' market in Corsicana. My husband cooks a lot, and he has very discriminating taste. He was always critical of what the children and I cooked, and we found that if we readjusted our recipes as he suggested, we placed at the fair. He is our official taster. He's from Louisiana, so he knows gumbo. I'll tell you the recipe, just like I do it at the fair for people watching. They get little tastes, and they seem to enjoy it."

LILLIE CROWLEY'S GUMBO

"You cut up okra, tomatoes, red and green sweet bell peppers, and you add fresh corn cut off the cob. Cut up an onion or two, and a few cloves of garlic—you've got to have garlic in gumbo.

"Sauté the onions, peppers and garlic in a little oil and butter. Add the tomatoes, okra and corn, and cook the vegetables until tender. Add seasoning salt, a little bit of sugar, and you can add ground or cut-up hot pepper. I do not add water; let the vegetables make their own juice.

"Serve the gumbo over a bed of rice. Cook the rice according to the package instructions, but when you put it on to cook, add a little oil or butter to keep it from sticking. I have found that the new yellow rice is wonderful.

"If you have leftover shrimp, chicken or meat, add that too, and heat it through for five minutes. Fresh parsley may be added just before serving."

LILLIE'S SQUASH DINNER IN A SKILLET

"You can use any tender summer squash–zucchini, crookneck, yellow, banana or pattypan, or a combination. The number of squashes depends on the number of people."

"Cut up an onion, and sauté it in oil and butter. Add the squash, sliced or cubed. Put in a light sprinkling of caraway seed, believe it or not, just to enhance the flavor. Allow one medium potato per person, cut up like the squash.

"If you have smoked sausage or ham, cut that up and add it to squash and onions, a clove or two of garlic if you wish, then seasoning salt. Very lightly, because the sausage will add the real flavor. Cover with a tight lid; the squash will make the liquid. But if it's too dry, add a little chicken broth, cover again and 'smother' until the potatoes are done. Serve with rye bread. A total dinner in a pot!"

ANASAZI STEW

Each year Elizabeth Peabody and her staff put out the State Fair of Texas
Prize Winning Recipes, *a collection of the previous year's winners. This recipe is adapted from the 1991 edition.*

Ray Calhoun of Richardson, Texas, a frequent winner, entered his stew in the Ethnic Foods division. He says, "Anasazi beans are an Indian variety from Colorado. They must be soaked and cooked in lightly salted water. Uncooked, they look like speckled limas, but are shaped like, and are similar to, pinto beans. Venison is commercially available from game ranches."

SERVES 4 TO 6

2 tablespoons vegetable oil	Water
1 pound venison, cut into 1-inch pieces (or substitute beef)	2 ears corn, roasted and cut from cob (about 1 cup kernels)
½ cup scallions (green onions), chopped	8 ounces anasazi beans, cooked (or substitute pinto beans)
2 Anaheim peppers (red or green), seeded and chopped	1 tablespoon cornstarch mixed with
Salt and pepper to taste	⅓ cup water (optional)

1. In a large heavy soup pot, heat the oil to sizzling. Brown the meat, turning often. Add the scallion and peppers, salt and pepper, and water to cover. Cook, partially covered, until the meat is tender.
2. Add the corn and cooked beans. Thicken with the cornstarch mixture, if you like.

BLANCO COUNTY FAIR & RODEO

Johnson City, Texas

FAIR DATE: THIRD WEEK IN AUGUST, FOR 3 DAYS, THURSDAY, FRIDAY AND SATURDAY

GAZETTEER: *Johnson City, the "Home of LBJ," is in the heart of the beautiful Hill Country, sixty miles straight north of San Antonio. The town is near several sites commemorating President Lyndon Baines Johnson that are open to visitors. The fairgrounds are on Highway 281.*

"We have a very rustic fairground," says Mary Earney, one of the fair's directors, "and we have such an old-fashioned fair. We have big dances on Friday and Saturday nights, and the rodeo is very important."

"The fair started in the Depression," says Ava J. Cox, "when the county agent came to educate farmers and their wives. Taught them how to cold pack and pressure can, and to make more with what they had.

"My grandfather, Sammy Lee Johnson, settled this town, and I am his oldest grandchild. Lyndon was my first cousin. He always supported the fair. I rode six miles to school on horseback, opening seven gates on the way, in order to get an education. I taught school for forty years, and I know a lot of people. Much of the Johnson ranch is now a national park. I have four acres of the original property. We used to take the cattle from Johnson City to San Antonio. We had a covered wagon and a bedroll; wasn't any hotel. Men and women were brought up to do ranch and farm work, no such thing as 'I can't.' We learned to cull chickens, to tell which were layers and breeders. We learned to judge a horse and a cow. I still ride my Brahma bull, Redman. My Brahmas are tame, 'cause they know I love them and feed them. Treat animals as you'd like to be treated, and they in turn will respond.

"I've always exhibited at the fair. The year that recycling was the theme, my project was a quilt made of old neckties. All I had to do was ask, and the ties came in like nobody's business. We cooperate in our community. I'm kicking at the door of ninety, but dates don't mean a thing."

"Blanco County Fair is very small," says Crystal Sultemeier. "The first night is the junior rodeo, from tiny barely walking ones to high school seniors. The small ones have stick-horse races. I used to compete in the rodeo, barrel racing. I live on a seventy-eight-hundred-acre ranch, and riding comes naturally to me. I've been in 4-H since the third grade. Some of my projects are food and nutrition, textiles and clothing, and swine and angora goats. I win blue ribbons all the time. Modest, aren't I? I love speaking in front of people and giving readings of fiction and poetry. I like the stuff that brings tears to your eyes. I love basketball and volleyball, and I'm manager for the junior high school football team. I'm president of our 4-H club and president of the Exchange Club; that's kids who every summer take trips to other states and stay with families. One year we went to Montana and stayed a week. I loved it; meeting new and interesting people. The cool weather was nice, coming from a hot Texas summer.

"Miss Ava Cox taught Mom in school, and she and my Aunt Josie are close friends. I call her Aunt Ava. She's a great woman—spins wool on a wheel and demonstrates it at school. In Johnson City, everybody is kin to everybody."

"They like to be noticed," says Ava Johnson Cox, here with a friendly Brahma.

AVA J. COX'S HOT WATER CORNBREAD

"Be sure to grease the top well," says Mrs. Cox, "and bake until brown around the edges. I prefer yellow cornmeal, but white is just as good. This was all they had to eat during the Civil War and afterward. The recipe has a lot of history behind it."

MAKES 1 PAN CORNBREAD

4 cups water

1½ cups cornmeal, yellow or white

½ to 1 teaspoon salt

3 tablespoons melted bacon grease or shortening, or vegetable oil

- Preheat the oven to 400 degrees.
1. Bring the water to a rolling boil. Sift in the cornmeal and salt, stirring constantly. Take off the heat. If any lumps form, beat them well with a wooden spoon or whisk to get rid of them.
2. Spread an 8-inch-square baking dish with one tablespoon of fat, swirling it up the side. Spread the batter in the dish, smoothing it. Cover it with the remaining fat.
3. Bake the cornbread for 1 hour, or until brown and crisp around the edges. Cut into serving pieces.

CRYSTAL'S HOBO'S SUPPER

"I often cook for the family," says Crystal. "Multiply this recipe by how many you are feeding."

MAKES 1 MEAL

Salt and pepper to taste
3 large onion slices, ⅓ inch thick
3 large potato slices, ⅓ inch thick

1 carrot, cut into slices ⅓ inch thick (optional)
Large hamburger patty

• Preheat the oven to 325 degrees.
1. For each serving, oil a 1-foot-square sheet of aluminum foil. Salt and pepper the onion and potato slices, and lay them in the center of the foil. Add the carrot slices, if wanted. Salt and pepper the meat, and lay it on top of the vegetables. Enclose tightly, and bake for 1½ to 2 hours. Do not open the foil until ready to eat.

Crystal Sultemeier.

TEXAS-BRAND BANANA BREAD

"I won a prize with this recipe," says Crystal Sultemeier. "I really love this bread."

MAKES 2 LOAVES

1½ cups sugar
½ cup vegetable shortening
2 eggs
½ teaspoon salt
1 teaspoon baking soda
¼ cup buttermilk

2 cups all-purpose flour
1½ cups mashed ripe bananas (about 3 bananas)
1 teaspoon vanilla extract
1 cup chopped pecans

• Preheat the oven to 350 degrees.
1. Spray two 9 × 5 × 3-inch loaf pans with vegetable spray.
2. In a large bowl, cream the sugar and shortening together until light and fluffy. Beat in the eggs and salt.
3. Stir the baking soda into the buttermilk. Add it and the flour to the mixture, and beat well. Mix in the bananas, vanilla and pecans.
4. Bake in the center of the oven for 45 to 60 minutes, until a toothpick inserted in the center of a loaf comes out clean. Cool in pans 10 minutes, then turn out to cool on a rack. Each loaf yields 16 half-inch slices.

PANHANDLE SOUTH PLAINS FAIR

Lubbock, Texas

FAIR DATE: THIRD SATURDAY AFTER LABOR DAY, FOR 8 DAYS

GAZETTEER: *The South Plains of Texas is a narrow strip of farming, cotton and—surprisingly—wine vineyard country. Lubbock is the major city, and the birthplace of rock 'n roll legend Buddy Holly. Fairgrounds are located off I-27.*

"We're the granddaddy of West Texas fairs," says Steve Lewis, the fair manager. "We pride ourselves on the number of organizations who support most of their work by what they make at the fair. Thirty-eight civic, fraternal and religious nonprofit groups participate. The Shriners have a bingo concession, the Foursquare church serves pancakes, and the Jaycees have a haunted house.

"Of course we have big-time entertainers. Elvis himself was here in 1956! I guess you know that Texas is twirling country. We have twirlers from all over West Texas, Oklahoma and new Mexico, with strutting and twirling competitions. The last Saturday of the fair around three hundred and sixty contestants perform. We have band competitions, too. A parade of bands at ten a.m. kicks off the fair; they're led by majorettes, and a color guard with fair officials in vehicles behind them. The march goes from downtown Lubbock to the fairgrounds, with ten thousand people watching along the route. We have thirty-five high school area bands, from very large to small schools. The bands compete in their own divisions, for trophies and for glory."

In the Sheep Barn.

Onedia Hagens is one of the assistant superintendents in the culinary department. "Our family has been involved in the fair for many years; four generations won blue ribbons one year, in cooking and in sewing. I won in cakes, my mother in crocheting, one daughter in cakes, and two sets of grandchildren in cookies. My daughters Kathy and Karen work at the fair, but I've yet to convince Debra. We are a farm family. Weather here can be fierce. We had twenty-two acres of vineyards, all destroyed by hail. Now we have replanted four hundred and fifty vines."

ONEDIA HAGENS' ENCHILADAS

"My three daughters and their husbands insisted that I share this recipe. It's a family favorite that I make for my husband Loyd's birthday. These enchiladas are mild, but you can make them hotter to suit your own taste."

MAKES 20 TO 24 ENCHILADAS

2 pounds relatively lean ground beef, preferably chuck

2½ cups chopped onion

1 tablespoon salt

1 teaspoon freshly ground black pepper

3 cups water

4 tablespoons (½ stick) butter

2 bell peppers, seeded and chopped

2 cloves garlic, minced

¼ cup flour

2 tablespoons chili powder

2 cups chopped stewed tomatoes, with juice

20 to 40 corn tortillas

   Vegetable oil for frying

1 pound grated Velveeta cheese

   Chopped scallions (green onions), including some of the green tops

   Chopped hot peppers, a combination of cherry peppers and jalapeños (optional)

1. First, cook the meat. In a large heavy pot, place the meat, 1/2 cup of onion, 2 teaspoons of the salt and the black pepper. Add the water. Bring to the boil, lower the heat and simmer for about 20 minutes. Remove from the heat and let the juices rise.

2. In a large heavy soup pot or Dutch oven, melt the butter over moderate heat. Cook the remaining 2 cups of onion, the peppers and garlic for about 6 minutes, stirring constantly. Stir in the flour, the remaining teaspoon of salt and the chili powder. Stir for 2 minutes. Add the tomatoes.

3. Draw off two cups of the meat stock from the meat pot and add to the chili mixture. (The meat should now be rather dry. If not, cook it down for a few minutes.) Set the meat aside. Simmer the chili mixture, uncovered, until thick, about 40 minutes, stirring often.

4. Lightly fry the corn tortillas in oil just until they turn limp. Dip them into the sauce, and fill with about 3 tablespoons meat and a sprinkling of grated cheese. Roll up, arrange on a warmed platter and top with remaining sauce, meat and cheese. Add a sprinkling of scallions and pass the hot peppers, if wanted.

FIVE

THE GREAT LAKES

Escanaba • **Michigan**

Shawano •

Plymouth • • Travers City

Wisconsin

• Kalamazoo

Sandwich • • Goshen • Delaware • Canfield
• **Ohio**
• Peoria • Columbus
Indiana Springfield

Illinois

Ohio, Indiana,
Michigan, Wisconsin,
Illinois

OHIO STATE FAIR
Columbus, Ohio

FAIR DATE: EARLY AUGUST, FOR 17 DAYS

GAZETTEER: *The fairground is next to I-71, where it intersects Highway 70. Columbus is the capital, and the site of Ohio State University.*

For an agricultural and entertainment extravaganza, come to the Ohio State Fair! But bring along your walking shoes and a game plan; there are seventy-seven thousand exhibits on the three-hundred-and-sixty-acre fairground. There's the largest state fair all-breed horse show in the Coliseum, from tiny Shetland ponies to massive six-horse hitches. There is a championship rodeo, and a daily parade. One building is devoted to the products of Ohio's eighty-eight counties, from eggs to aerospace.

"I've been coming to the fair since I was two years old," says John Evans, retired fair manager. "And that was in 1920. My father showed champion shorthorn cattle. As a kid, what I liked were the tractors and threshing machines. Columbus was a national outlet for farm equipment, and all the manufacturers, like John Deere, exhibited at the fair. I'd collect all their handouts in a paper bag, and pore over them in the winter months. My hobby now is antique farm equipment.

"In 1937 I won with our grand champion shorthorn. I had the hide tanned, and a robe made out of it. When I was president of the Old-Timers' Club—that's for people who have attended sometime within fifty years—I recalled that it was exactly fifty years ago I'd won with the steer."

"We're a fair family," says Katie Grimm. "My husband, Fred, was a County Extension Agent, and he helped start our local Ottawa County Fair about twenty-five years ago. We both worked closely with 4-H'ers, and sent those who had won locally to the state fair. My son, David Grimm, is manager of the Champlain Valley Exposition in Vermont.

The Ford Classic Truck and Tractor Pull.

Dressed alike for the Guys and Gals Sheep Lead Competition.

"Back in 1971, I went to the All Ohio Bake-A-Rama at the Ohio State Fair, and I won! You could make two pies, but the good Lord helped me, and I only made one. I was the first one finished. It was a two-crust pie flavored with cinnamon and nutmeg. I had to use soft summer apples instead of fall ones, and I made the crust with lard. One judge said, 'How can you be so calm?' I told him, 'With a husband like Fred, and two boys, I have to be!'

"What I enjoyed most at the fair were the dairy exhibits. An artist in a chilled sealed unit sculpted a cow out of butter. I thought that was fabulous. They had the best ice cream there, too."

Best in Ohio!

KATHERINE GRIMM'S TOMATO SALSA

"After he retired, my husband, Fred, had a television show in Toledo called 'Fresh Directions,' all about fruits and vegetables. I cooked all the food, and he took it over to the station. I enjoyed it, because we both love fresh produce. At our age I can't believe that we like Mexican food, but we do! It may be hot in the mouth, but not in the stomach,"
says Katie.

MAKES ABOUT 4 CUPS

10 to 12 fresh green jalapeño peppers, seeded and finely chopped
⅓ cup finely chopped parsley
⅓ cup chopped onion
3 large garlic cloves, finely chopped

3 to 4 fresh tomatoes, chopped
2 cups drained tomatoes, chopped
1 teaspoon chopped fresh oregano, or ½ teaspoon dried, crumbled
Salt and pepper to taste

• Combine all the ingredients. Serve with taco chips, eggs or cold meats.

CANFIELD FAIR
Canfield, Ohio

FAIR DATE: THURSDAY BEFORE LABOR DAY, 5 DAYS

GAZETTEER: *"Canfield" is a nickname; The Mahoning County Agricultural Society Fair is the true name. In the rolling hills of rural Mahoning County, the major crops are apples and strawberries. Youngstown is the largest city. The fairground is one mile south of the I-224 and Route 46 intersection.*

Ohio's largest county fair has a long, rich history. The year 1851 was especially full of wondrous sights. Two hundred oxen, in two trains of fifty yokes each, impressed the crowd, as did a bull trained to harness. A diploma was awarded to Ira Jones for his exhibit of a patented shower bath. The Poland Sons of Temperance displayed a banner advocating their cause, presented to them by the ladies of the town of Poland. Ten years later, during the Civil War, army volunteers home on furlough helped eat the prize-winning cakes on the last day of the fair.

Today's fair pleases both city and country folk. "Youngstown was a steel town, so it's fitting that we have the Draft Horse and Draft Pony Pig Iron Derbies," says Kathy Bennett, one of seventeen fair board members (and the first woman member in the fair's hundred and fifty years). "The animals pull sleds weighted with pig iron, which is a by-product of steel making. We also have all the agricultural judging that anyone could want, including some breeds of cow that not every fair shows. We have Brown Swiss and Ayrshires, and milking Shorthorns.

"'Something to Crow About!' is our fair's motto," says Kathy. "Our mascot is a hundred-pound, eight-foot-tall fiberglass rooster. He was kidnapped back in 1975 by high school kids, and ever since then he's been put away annually for safekeeping. We also have rooster crowing contests, in junior and senior divisions. You bring in your rooster, and the winners are the ones that crow most often in a given space of time."

A DINNER TO CROW ABOUT

Fairgoers at Canfield, two hundred and fifty at a time, sit down to chicken prepared by members of the Ruritan Club, a rural service organization. A flavorful chicken half is served with coleslaw, applesauce, potato chips and a soft drink. The Ruritans estimate that since 1960 they've served over a quarter of a million dinners, with all the money made going back into the community.

GREENFORD RURITAN CHICKEN AT THE FAIR

The Ruritans place chicken halves on racks over a glowing charcoal pit. The chicken is turned continually from one end of the pit to another, and frequently sprayed with basting sauce from a garden hose. Cooking in this manner takes about two hours. According to a Ruritan spokesman, "This amount of sauce takes care of three hundred to seven hundred chicken halves, depending on who's cooking. We're all volunteers, you know."

MAKES ABOUT 3 1/4 GALLONS SAUCE

1½ gallons water	1 cup vegetable oil
1½ gallons cider vinegar	1 10-ounce bottle Worcestershire sauce
1 cup salt	

- Place the chickens on racks over glowing coals. Spray and turn several times until browned and tender.

GREENFORD RURITAN CHICKEN AT HOME

Oven frying simplifies the preparation and cuts down on the charcoal broiling time.

SERVES 6

6 chicken halves	5 teaspoons salt
½ cup water	2 teaspoons vegetable oil
½ cup cider vinegar	1 tablespoon Worcestershire sauce

- Preheat the oven to 350 degrees.
1. Place the chicken halves on a rack in a large pan. Bake until an instant-read thermometer placed in the thickest part of the leg registers 165 degrees. Remove the chicken, drain off the pan juices, and arrange the chicken in the pan in one layer.
2. Bring all the remaining ingredients to the boil. Pour over the chicken. When cooled slightly, refrigerate overnight. Turn occasionally
3. When ready to finish the cooking, heat coals to glowing. Place the drained chicken on a rack and cook just long enough to brown nicely.

CLARK COUNTY FAIR
Springfield, Ohio

FAIR DATE: LAST WEEK OF JULY, FOR 7 DAYS

GAZETTEER: *Agriculture is the leading industry in this pretty part of southwestern Ohio, located thirty miles from Dayton and Columbus. For the fairgrounds, take exit 59 off I-70.*

The present fair, begun in 1948 as an agricultural showcase, has evolved into a week-long event on one hundred and twenty-seven acres, "second only to the Ohio State Fair," says William Schwartz, executive director of the Clark County Agricultural Society. "We think we do a good job of blending the rural and the urban. Ours is one of the larger junior agricultural fairs, and we have a Junior Fair board as well as our Senior Fair board.

"A rabbit club is part of the Junior Fair. The kids have a queen and court—rabbit royalty! Winning is based on knowledge and care. The court also has dukes and duchesses.

"We've had a knack over the years of booking entertainment talent just before they become known. We get them just as they are 'hitting.' A few years back we booked an unknown group called New Kids on the Block. By fair time, they were the number one idols. We don't charge for entertainment, and it's first come, first served. The day they were to appear we had ten thousand kids standing in line at nine a.m. They stood in the hot sun all day for the nighttime performance. We had to close the Youth Building and open it as a first aid station; they were passing out from the heat. That year we registered our highest Monday attendance, and our lowest vendor income. The kids wouldn't get out of line to buy food or drink, and the adults stayed away, knowing it would be chaos. It kinda backfired on us."

ETHEL WADDLE'S CREAM CHEESE POUND CAKE

Mrs. Wilbur Waddle is a champion baker, and many-time winner at the Clark County Fair. The cake freezes beautifully, she says.

MAKES 1 CAKE

2 sticks butter, at room temperature
1 (8-ounce) package cream cheese, at room temperature
3 cups sugar

6 eggs
3 cups sifted cake flour
2 teaspoons vanilla extract

1. Combine the butter and cream cheese, preferably in the bowl of a heavy-duty mixer. Beat on low speed until well mixed. Add the sugar gradually, and beat until light and fluffy.
2. Beat in the eggs, one at a time, beating after each addition. Beat until the mixture is light. Add the flour and vanilla, and beat only until the flour disappears.
3. Pour the mixture into a greased 10-inch pan. Place in a *cold* oven, and turn the thermostat to 325 degrees. Bake the cake for 1 hour and 10 minutes. Cool on cake racks, and chill before serving.

THE 4-H FARMERETTES
AND THE HEALTHIEST GIRL IN CLARK COUNTY

Dorothy Drake, health contest winner.

"I'd say that the 1950s were perhaps the height of the 4-H movement," says Dorothy Drake, whose family has farmed in Clark County since 1811. "After school, 4-H was our major focus, and the center of our social life. We met once a week.

"However, everything seemed to be for the boys when it came to caring for animals. I didn't want to bake pies and do those girl things; I liked animals. I fed sheep and Black Angus steers, and wanted to take them to the Clark County Fair. Other girls felt like me, and so my friend Carolyn and I organized the Farmerettes, the Oak Grove 4-H livestock club. My little sister Sarah was our mascot. We'd help each other, like washing our lambs before the fair. Wonderful people named Margaret and William Welsheimer, Carolyn's parents, encouraged me. I grew up with their son Philip too, who is actually farming our place now."

Dorothy won a number of blue ribbons with her sheep, cattle and sewing, but she also received another honor: twice she was the Health Winner at the Clark County Fair. "I recall you had to have a doctor's certificate and good teeth," Dorothy says. "You also had to have a good record in 4-H projects and community service. The 'health' stood for one part of 4-H's emblem, 'Head, Heart, Hands and Health,' I won when I was fourteen, and again when I was seventeen. My sisters, both beauty queens, got a big laugh out of 'the healthiest girl in Clark County.'"

EVA HUPMAN'S RHUBARB PIE

When Dorothy Drake does bake a pie these days, it's apt to be neighbor Eva Hupman's superb rhubarb. Lard makes a wonderfully flaky pie crust. It may be out of medical fashion now, but Eva Hupman has lived for almost one hundred years making and eating lard pie crust, so it can't have done that much harm.

MAKES 1 PIE

THE FILLING:

1½   cups sugar   3   cups cubed pink rhubarb (1-inch cubes)

| 3 | tablespoons flour | A few drops of red food coloring (optional) |
| 1 | tablespoon cornstarch | 4 teaspoons butter |

THE PASTRY:

| 2 | cups sifted flour | ⅔ cup lard |
| 1 | teaspoon salt | 5 tablespoons water |

- Preheat the oven to 425 degrees.
1. Mix all the filling ingredients except the butter together. Let stand while you make the pastry.
2. Mix the flour and salt. Cut in the lard with a pastry cutter, just until the mixture resembles coarse meal. Sprinkle the water over the mixture and combine with a fork. If it's a hot day, chill the pastry briefly in the freezer before you roll it out.
3. Roll out the bottom crust and line a 9-inch pie pan with it. Pile the filling in the pan and dot with the butter. Make a lattice topping and flute the edges.
4. Bake the pie in the center of the oven for 10 to 15 minutes. Reduce the heat to 350 degrees, and bake 30 to 40 minutes more, until the rhubarb is tender when pierced with a skewer.

A. B. GRAHAM, PIONEER OF THE 4-H MOVEMENT

In 1902, the school superintendent of Springfield Township, Clark County, Ohio, organized a group of students to conduct agricultural experiments. This program was a forerunner of the worldwide 4-H movement.

Graham's chief aim was to encourage boys to stay on the farm and to feel the rightful pride of the farmer's place in society. Education was an important part of that aim.

"He kept many a boy in school and made many a child's life better," says centenarian Jennie A. Irie, who was a nine-year-old member of that pioneer class of 1902, as was her late husband. "He gave children self-worth. Families fell apart when they went to the city; they stayed together on the farm. 'Reach for goals, and make the best of yourself,' he said. I remember his beautiful handwriting on the blackboard, and how fastidious he was, in his suit and tie.

"Mr. Graham showed us the different hays and grasses, and different soils. He had a vision, like Thomas Jefferson."

A. B. Graham, pioneer of boys' and girls' agricultural clubs.

In 1905, Graham became the first Extension Superintendent at Ohio State University, his alma mater. Among his goals were:

1. To elevate the standard of living in rural communities.
2. To emphasize the importance of hard work and habits of industry, which are essential to building strong character.
3. To acquaint boys and girls with their environment and to interest them in making their own investigations.
4. To give the boys, who shall become interested in farm work, an elementary knowledge of agriculture and farm practices and girls the simplest facts of domestic economy.

Graham went on to work for the Department of Agriculture, leading the teaching of extension specialists throughout the country. In 1952, the golden jubilee of the founding of 4-H, the government issued a commemorative stamp bearing the 4-H motto, "To Make the Best Better."

JENNIE IRIE'S NOODLES

"My mother was always a good plain cook," says her daughter Janet. "Mother says to cook a lean roast of meat and remember that you want plenty of broth for the noodles. Chicken and chicken broth may also be used. To measure out the extra water that may be needed, she uses the expression 'take a half-eggshell of water.'"

MAKES 10 OUNCES NOODLES (ABOUT 4 CUPS COOKED)

2 cups all-purpose flour
½ teaspoon salt

2 large egg yolks, slightly beaten
½ cup water

1. Sift the flour before measuring. Add the salt and sift again. Combine the egg yolks and water and stir into the flour mixture. If the dough seems stiff or too dry to handle, add a few drops of water at a time, but only if absolutely necessary.
2. Turn out onto a floured board and knead for about 5 minutes, until well blended. Cover with plastic wrap and let stand at room temperature for about 30 minutes.
3. Divide the dough into 4 parts and roll out each part on a floured surface until almost paper thin. Roll each circle up like a jelly roll and slice evenly into narrow strips. Let the noodles dry at room temperature. Store in a cool place until ready to use.
4. Drop into rapidly boiling broth or water and cook until tender.

DELAWARE COUNTY FAIR

Delaware, Ohio

FAIR DATE: THIRD WEEK IN SEPTEMBER, 7 DAYS, SATURDAY THROUGH FRIDAY

GAZETTEER: *Delaware, the site of Ohio Wesleyan University, is about twenty miles north of Columbus. Take Route 23 to the fairground.*

"Our fair is home to the harness race pacers' classic, a race called 'The Little Brown Jug,'" says Bill Lowe, fair manager. "It was first raced in 1946, and it's now the best attended harness race in North America, perhaps the world—I'm not sure, but Australia may have a bigger race. The atmosphere of a county fair and a great race makes for a great blend—it's a throwback, on a grand scale, to the days when small farmers raced each other. We have harness racing on five days of the fair."

"We've got the largest all-horse parade east of the Mississippi," says Diana Winter, parade chairman. "It's a circular parade, two and three-tenths miles from the fairground and back. It's easier for the teamsters to pack and unload in the same place. We have horses, ponies and mules—almost four hundred animals. We've got equestrian units from miles away, a number of floats, and the Heinz eight-horse black Percheron hitch. People along the parade route hold barbecues and porch parties and set up tents in side yards. They just party all afternoon as the parade goes by!"

"Ours is a traditional country fair with rides, livestock exhibits, and of course the harness racing," says Bill Lowe. "I was told by a man in the carnival business that there are two things that bring in the crowds—sex and motors. Now, we can't have sex at the county fair, and we can't do demolition derbies, tractor pulls or motorcycle racing because we have to protect our racetrack. Even so, we must be doing something right. We have a great time and we do get the crowds."

Ray Remmen driving Beach Towel, 1990 winner of the Little Brown Jug.

HELEN THOMPSON'S REUBEN SPREAD

Helen is an avid race fan. Her husband, Tom, is president of the Brown Jug Society, the body that runs the races. "I don't cook much," says Helen, "but I can heartily recommend this hors d'oeuvre. Serve it with those little thin rye bread slices."

SERVES 10

8 ounces Swiss cheese, grated

8 ounces sharp cheddar cheese, grated

1 (14-ounce) can sauerkraut, drained

1 (4-ounce) package corned beef, torn into small pieces

¾ cup Hellmann's or Best Foods mayonnaise

- Preheat the oven to 350 degrees.
- Mix all the ingredients well. Put into a small casserole or 8-inch-square Pyrex dish. Bake, covered, 30 minutes.

THE TRACTOR PULL. LET'S GO-O-O!

Tractor and truck pulls are a major, major draw at country fairgrounds throughout the United States; only Alaska and Hawaii, to date, are pull-less. In principle, the sport is simple. These events are the motorized descendants of draft horse pulls, where huge, burly animals drag stone sleds loaded with increasingly heavy weights for a prescribed distance. The strongest horse wins.

The mightiest tractor or truck, competing in divisions where each vehicle weighs the same, wins when it pulls a load the longest distance down a three-hundred-foot track. How a vehicle becomes mightiest is the essence of the sport; it takes great engineering sophistication to tuck up to fifteen hundred horsepower under a hood. These vehicles may look like ordinary trucks or field tractors, but turbo charging, extra torque, lightweight metals and mechanical ingenuity separate the winners from the also-pulls.

The invention of the weight transfer sled changed the sport from a by-guess-and-by-golly, is-Joe-dragging-the-same-weight-as-me? contest into one where all competitors are equal. Briefly, the vehicle is hooked to a sled that has a weight box over its rear axle. As the vehicle moves down the track, the weight on the sled slides forward on inclined rails until the bottom of the sled is in contact with the ground. Friction slows, then stops, the pulling vehicle. If you make it the whole three hundred feet, that's called a full pull. You have won, or you must pull off–with added weight–against equal competitors until someone eventually wins. Let's go!

Doc Riley, communications director of World Pulling International and the National Tractor Pulling Association, says that his group oversees safety–pullers wear helmets and flame-retardant clothing and have fire extinguishers on board–and sets rules for pulling events. A good example of a committed puller, he says, is Alan Snead, who owns Snead's Restaurant in Lancaster, Ohio, with his wife Rosie:

"I pull in the truck category, and I drive a 1951 Chevrolet pickup truck. We go all over–to the Hamburg County Fair, and to events in Saskatchewan, in Minnesota, wherever.

"We take the truck inside a forty-foot motor home transporter. The motor home is self-contained for living, even has TV. Behind it we tow a Chevy van for going to town.

"A bunch of us travel together, get to events a day early, go to town and buy food. We barbecue and have a good time. We keep it simple–we're not in the business of doing dishes, and we have a lot of 'bench racing' to do (essentially, that's B.S.), and cooking would take too much time away. We talk about changes, about what went right, went wrong.

"At a three-day event you'll have two passes down a track, a grand total of about forty seconds running your vehicle. In up to sixty hours, you have a lot of time to bench race and work on your vehicle, if needed."

Alan Snead's FWO pulling truck, named "Lost in the 50s."

ROSIE SNEAD'S LASAGNE

Alan and his wife Rosie own Snead's Restaurant, a moderately priced, full-service establishment in Lancaster, Ohio. Rosie cooks.

SERVES 10

1 pound sweet Italian sausage (or sweet and hot mixed), removed from casings
½ cup chopped onion
2 cloves minced garlic
1 (16-ounce) can tomatoes
1 (8-ounce) can tomato sauce
1 (6-ounce) can tomato paste
2 teaspoons crumbled dried basil
1 teaspoon salt

8 ounces lasagne noodles
1 tablespoon vegetable oil
2 eggs
2½ cups ricotta or cream-style cottage cheese
¾ cup grated Parmesan or Romano cheese
3 tablespoons chopped parsley
1 pound mozzarella cheese, thinly sliced

- Preheat the oven to 375 degrees.
1. In a large heavy pot over moderate heat, cook the sausage, onion and garlic until the meat is brown, stirring to break up the lumps. Drain off the fat. Stir in the tomatoes, undrained, and break them up. Stir in the tomato sauce, tomato paste, basil and salt. Bring to the boil, cover the pot, lower the heat, and simmer for 15 minutes, stirring often.
2. Meanwhile, cook the noodles in a large quantity of boiling salted water, to which you have added the oil, until tender; fish out a noodle and test after 8 minutes. Drain the noodles in a colander and rinse them under cold water. Set aside.
3. In a large bowl, beat the eggs. Stir in the ricotta, 1/2 cup of the grated cheese, the parsley, salt and pepper. Oil a 13 × 9 × 2-inch baking dish. Spread half the noodles evenly in the dish; follow with half of the egg-cheese mixture; half of the mozzarella, then half of the meat sauce. Repeat the layers. Sprinkle the remaining grated cheese over the top.
4. Bake for 35 minutes, or until bubbling. Let stand 10 minutes to settle.

ELKHART COUNTY 4-H FAIR

Goshen, Indiana

FAIR DATE: END OF JULY, 9 DAYS

GAZETTEER: *Goshen is in northern Indiana, between South Bend and Fort Wayne. Fairground is just off U.S. 33.*

"Years ago the fair board decided that we would provide a showcase for our young people in 4-H," says Dorothy Kerchner, the fair's executive secretary. "But our motto is 'Something for Everyone,' so we have a Senior Queen, who reigns over the Senior Citizens' Days at the fair and participates in all activities. The winner is always someone over sixty who has done a lot for the community."

The traditional 4-H classes are of course well represented—showing dairy cows and goats, rabbits and sheep, and preserving food—but kids can also compete for ribbons in categories that didn't even exist when 4-H was founded. There's model plane building, microwave cooking, and rocket and electrical circuit board build-

ing, for example. 4-H'ers can also opt to wrap gifts, ice cakes, and publish the *Clover Chronicles*, the Media Club's newspaper, for six days of the fair.

"The biggest entertainment attraction is the Tractor Pull, which goes on from eleven a.m. until midnight," says Dorothy. "People get a seat by nine a.m. and stay all day." Important, too, is the Power Pedal Pull, in three categories for kids aged five to ten, where contestants pull weights attached to small pedaled tractors. "No standing up to pedal," the rules say.

Adults can join in events for not-so-serious fun, like the Sweet Corn Eating Contest—no butter or salt, and cooled for

Maze built of 300 bales of straw.

the kids. An elderly man won the Longest Apple Peel contest three years in a row. His best: a peel one hundred six and a half inches long, in one piece.

"We are the dairy capital of the state," says Mary Ann Lienhart-Cross, Extension educator, consumer family science coordinator and adviser for many fair activities. "So we're proud of the Great Homemade Ice Cream 'Crankoff.' You can compete as an individual or as a team. Some families really get into it—they come wearing their own printed team aprons."

FRENCH SILK ICE CREAM

The Parcell family, Deb, Dave, Becky and Jimmy, won first prize in 1992 at Elkhart with this incredibly smooth ice cream.

MAKES ABOUT 2 QUARTS

6 egg yolks, beaten	1 cup semisweet chocolate chips
1 cup sugar	1 tablespoon vanilla extract
2 cups half-and-half	3 cups heavy or whipping cream

1. In a large saucepan, combine the yolks, sugar, half-and-half and chocolate chips. Heat gently, stirring often, until the mixture is bubbling. Off the heat stir in the vanilla and cream.
2. Chill the mixture well. Freeze in a 1-gallon ice cream freezer, following manufacturer's directions.

RPM PATTIES

"We promote beef and pork in Elkhart County," says Mary Ann Lienhart-Cross.
"Three of us got together to come up with this recipe, which won the Pork Cookoff here and
at the Indiana State Fair, too. We named it for ourselves: Roger, Patty, and Mary Ann.
Don't overcook, as today's pork is lean, and you don't want it to dry out."

SERVES 3

1	pound ground pork	1	teaspoon freshly ground pepper
2	tablespoons sugar	1	teaspoon ground cinnamon
2	tablespoons soy sauce	3	spiced apple rings, diced
1	tablespoon grated horseradish	3	spiced apple rings, whole

1. Mix together all the ingredients except the whole spiced apple rings. Mix lightly but thoroughly; do not compress the meat too much or the patties will be tough.
2. Shape the mixture into 6 patties. Place each apple ring between two of the patties and pinch together to enclose the ring.
3. Have ready an outdoor fire, the coals covered with ash. Place the patties on foil on the grill so that they cook just enough to firm and hold their shape, then place them directly on the grill rack. Cook, turning once, until done; 160 degrees on an instant-read thermometer. (The patties may also be broiled or fried.)

NORTHWESTERN MICHIGAN FAIR

Travers City, Michigan

FAIR DATE: SECOND WEEK OF AUGUST, 7 DAYS

GAZETEER: *Travers City, surrounded by lakes, is a prime vacation area. In winter there is cross-country skiing and snowmobiling; in summer the population can swell to as much as one hundred thousand. The fairgrounds are off M-37, 10 miles from Travers City.*

This 1872 drawing by William Holdsworth, Jr., received a "best" in its category.

"Our present fair began in 1907," says Judy Mork, fair secretary, "and it was once called the Great Travers City Fair. We are a five-county fair, and we hold the fair in the peak of vacation time, so we get a lot of visitors.

"Area farmers like to compete. We have Farm Day to honor them, and we have a tribute to the 4-H. The Travers City Fair has a big draft horse show, with pulls, and we have harness racing."

"In 1992 I submitted a hundred and twenty-three projects—canned goods, afghans, crochet work, and baked goods—and I won fifty-seven blue ribbons," says Mary Keyes. "I was named Homemaker of the Year. I do projects on weekends and evenings. I work for Michigan Bell as a customer service representative now, but I used to manage the family farm for my in-laws. We raised angora goats, and sheared them twice a year. Their hair—mohair—is beautiful. Unfortunately, the wool price is way down. The price had gone from eight or nine dollars a pound to ninety-eight cents a pound by the time we sold the herd. I really miss going to the barn.

"I love these animals. They are stately looking; elegant, really. They have a nice temperament. Most goats have their horns removed, but with angoras, you look for good horn structure. Each herd has a dominant male, and as goats will fight for top place, you know that a fine rack of horns indicates a well-bred animal. We once had a bear attack; the bear came out of the swamp and killed a kid. He mauled the mother, too. We never caught the bear."

MARY KEYES' CHERRY JAM

"Travers City is the national cherry capital, and when we have a cherry festival, this town turns into chaos," says Mary Keyes. "Michigan is famous for its cherries, especially the sour ones. They make delicious jam. You can also make the jam with sweet cherries, but if you do, you must add lemon juice to counteract the sweetness. The little bit of butter added to the boiling fruit prevents a lot of frothing."

MAKES 6 CUPS

4 cups finely chopped (¼-inch) pitted cherries	5 cups sugar
¼ cup freshly squeezed lemon juice (for use only with sweet cherries)	1 box pectin, such as Sure-Jell
	½ teaspoon butter

1. Sterilize both jars and lids according to manufacturer's instructions.
2. Place the cherries in a large heavy nonaluminum pot. (If you are using sweet cherries, add the lemon juice.)
3. Measure the sugar into a large bowl, so that you can add it to the pot all at once. Do not use less sugar than called for! Set aside.
4. Add the pectin and the butter to the fruit. Bring the mixture to a full rolling boil over high heat, stirring constantly. Quickly add all the sugar and bring to the boil again, stirring constantly. Still stirring, boil for 1 minute. Off the heat, remove any froth with a slotted metal spoon.
5. Fill the hot jars at once to within 1/4 inch of the top. Wipe the threads and the jars. Cover quickly with lids, and screw tightly. Invert the jars for five minutes, then turn upright, or process for five minutes in a boiling water bath. After 1 hour, check the seals.

Winning jelly display.

SWEEPSTAKES PUMPKIN BREAD

"My bread won me the first-place ribbon for pumpkin bread, and the sweepstakes ribbon for all the breads, at the 1992 Northwestern Michigan Fair," says Mary Keyes.

MAKES 2 OR 3 LOAVES

⅔ cup solid vegetable shortening	½ teaspoon baking powder
2⅔ cups sugar	1½ teaspoons salt
4 eggs, lightly beaten	1 teaspoon ground cinnamon
1 (16-ounce) can pumpkin puree	1 teaspoon ground cloves
⅔ cup water	⅔ cup coarsely chopped nuts
3½ cups all-purpose flour	(walnuts are best)
2 teaspoons baking soda	⅔ cup raisins

- Preheat the oven to 350 degrees.
- Grease three 8½ × 4½ × 2½-inch loaf pans or two 9 × 5 × 3-inch loaf pans. Line the bottoms with wax paper cut to fit.
1. In a large bowl, cream the shortening and sugar together until light and fluffy. Stir in the eggs, pumpkin and water.
2. Whisk together the flour, baking soda, baking powder, salt, cinnamon, and cloves. Blend into the pumpkin mixture. Stir in the nuts and raisins.
3. Divide the batter among the pans. Bake about 70 minutes, or until a toothpick inserted in the center of a loaf comes out clean. Cool in the pans 10 minutes; turn out and finish cooling on cake racks.

COME TO THE FAIR

The sentiments and enthusiasms expressed in this paean to fairs are as relevant today as when these couplets appeared in the Travers City *Herald* on October 2, 1873. The author is unknown.

Attention, now, good people, pray!
A word or two I'd like to say
About the great Grand Traverse Fair,
For which of course you'll all prepare.
Though Barnum does go round and blow,
He can't come up to us, you know.
Elephants with trunks are fine
But not as useful, quite, as swine.
So bring from orchard, farm and field
The best the ground this year doth yield.
Bring from the work shop fruit of labor,
Come yourself and bring your neighbor.
Bring on your squashes, plums, potatoes,
Your biggest pumpkins and tomatoes.
Bring ducks and chicken, turkey, geese—
Bring anything you know will please.
Great cabbage-heads, and grapes and flowers—
The works of art and leisure hours—
Pictures, paintings, patch work, fruits,
Carpets, stoves and patent boots,
Cultivators, pigs and calves—
Never do a thing by halves.
Miss Jones, I think I heard it said
Will bring her nice salt-rising bread.
I really think 'twill stand the test
Though Mrs. Brown likes yeast bread best.
Miss R. the premium deserves
For those delicious plum preserves.
October is the month, you know,
For the prodigious union show,
We all will have a jolly time—
And here I'll end my simple rhyme!

KALAMAZOO COUNTY FAIR

Kalamazoo, Michigan

FAIR DATE: LATTER PART OF AUGUST, 7 DAYS, SUNDAY TO SUNDAY

GAZETTEER: *The fairgrounds are on the east side of Kalamazoo, just off I-94.*

"Originally Kalamazoo was agricultural, and still is to a great extent," says Garrett Ver Meulen, a longtime fair board member, whose four brothers and two sisters are also associated with the fair. "It's diversified— corn, beans, swine, dairy, sheep, goats and horses. If Kalamazoo doesn't have it, nobody has!

"Better than half of the livestock at the fair is 4-H. A lot of suburban children are in 4-H, and they show rabbits and chickens.

"We have a Demolition Derby, a Tractor Pull, and now motorcycle racing. We have five days of horse racing. We find we have to try new things all the time, because people want change, but they still want the draft horse pulling contest."

Mrs. Ver Meulen has spent over a decade as superintendent of Open Class Baking, and has "put in over forty years in 4-H." She's proud of Kalamazoo County's many fine bakers, both women and men.

ALL IN THE FAMILY

"I was born in Kalamazoo and have lived all my life in the county," says Betty Jean Williams.

"I began exhibiting at the fair in 1967, when I realized it was open to everyone. Until then I'd thought one had to be a farmer to enter.

"I was a 4-H leader for many years in our town of Scotts. My daughter Sheryl, my granddaughters Kelly and Jessica, and I have all won prizes at the county fair and the state fair for needlework. Kelly has also won trophies in the goat division, and Reserve Champion for her dairy beef feeder steer. Showing at the fairs—county and state—has become a standard of excellence in our family."

Goat and friend.

BETTY JEAN WILLIAMS' SEVEN-GRAIN BREAD

MAKES 2 STANDARD OR 4 SMALL LOAVES

1 package active dry yeast	1 cup milk
¼ cup warm water	4 tablespoons (½ stick) butter, melted
1 cup seven-grain cereal (found at	1½ teaspoons salt
health food stores)	3 tablespoons brown sugar
1 cup boiling water	5 cups unbleached flour

1. In a large bowl, sprinkle the yeast over the warm water and let stand. The mixture will become frothy.
2. Place the cereal in a bowl and pour the boiling water over it. Let stand until cooled.
3. Meanwhile, heat the milk to 110 degrees on an instant-read thermometer. Stir in the butter, salt and sugar. Pour the mixture over the cereal.
4. Stir the milk mixture into the cooled cereal, then stir this and the flour into the yeast. Knead well, at least 15 minutes (I find that the long kneading gives grain breads a finer texture). Turn into a greased bowl, rotate the dough so that all the surface is greased, and cover with oiled plastic wrap. Let rise until doubled, about 1 ½ hours.
5. Punch down and knead for 5 minutes. Cut into portions, shape into loaves and put into greased loaf pans. Let rise again, covered with oiled plastic wrap, until the dough has risen above the pan and is rounded, 30 to 45 minutes.
• Preheat the oven to 400 degrees.
6. Bake standard loaves for about 1 hour, small loaves for 40 minutes, or until browned on top and hollow sounding when thumped.

JACK'S APPLE PIE

Jack Jadkowski says, "I was introduced to baking in junior high school, when I had to take a required home economics course. My first reaction was 'Why?,' but as time went on I found baking fun. Some fifteen years ago my wife encouraged me to enter our county fair competition. Winning blue ribbons, and then in 1992 the rosette for best in show, are treasured moments."

MAKES 1 PIE

THE CRUST:

2 cups all-purpose flour	½ teaspoon salt
⅔ cup solid vegetable shortening	4 to 6 tablespoons very cold water

THE PIE FILLING:

6 to 8 apples, McIntosh or Granny Smith, peeled and thinly sliced	1 teaspoon ground cinnamon
¾ cup sugar	¼ generous teaspoon freshly ground nutmeg
3 tablespoons all-purpose flour	4 tablespoons (½ stick) butter, softened
⅛ teaspoon salt	

1. Make the crust: Combine the flour and salt, add the shortening, and cut it in with 2 butter knives until the mixture resembles coarse meal. Mix in water, a tablespoon at a time, until dough holds together. Wrap in plastic wrap and refrigerate while you make the filling.
2. To make the filling: Toss the apple slices with the sugar, flour, salt, cinnamon and nutmeg in a large bowl.
* Preheat the oven to 400 degrees.
3. Divide the dough in half and roll it out on a lightly floured surface. Arrange the bottom crust in a 9-inch pie pan and spread 2 tablespoons of the butter evenly over it. Add the pie filling and dot with the remaining butter. Cover with the top crust, crimp the edges, and cut several vents for steam to escape. Bake 50 to 60 minutes, until golden brown.

UPPER PENINSULA STATE FAIR

Escanaba, Michigan

FAIR DATE: THIRD WEEK OF AUGUST, FOR 6 DAYS, TUESDAY THROUGH SATURDAY

GAZETTEER: *Michigan's Upper Peninsula, lying between Lake Superior to the north and Lake Michigan to the south, is separate from the lower part of the state, and is connected only by the five-mile-long Mackinac Bridge. The U.P., as it is known, has agricultural and timber activity, with paper and lumber mills. The fairgrounds are off U.S. 2 and 41, two miles from the Escanaba business district.*

"The Mackinac Bridge really opened tourism in the U.P.," says fair manager Richard Ostrander, "and it means more visitors for us. We have Native American Day at the fair, with tepees set up, dancing, and crafts and food for sale. We have a number of tribes here, including the Chippewa and the Ojibwa.

"The Steam and Gas Village Engine Association constructed a village on the fairgrounds, with a museum depicting the history of agriculture and timber cutting. A shingle mill is part of the village, and the buildings were put up with lumber cut in the sawmill. There are old threshing machines and a rock crusher. All the equipment works.

"Quilting guilds are popular here, and the Quilt Block Contest draws a lot of entries. Contestants submit an eighteen-inch-square block, following a theme the fair board selects. One year it was all the animals—swine, horses, sheep, cows—that you see at the fair. Other themes have been floriculture and carousels. It takes twenty blocks to make a standard quilt, and if your block is chosen, you get a nice prize and recognition. We have experts; it takes quite an eye to arrange the blocks in a quilt, select the bor-

Cindy Clark, quilt coordinator, admiring the quilt commemorating Michigan's Sesquicentennial Celebration.

der, do the stitching and the tying. Sometimes on Governor's Day, the Governor will put in a few stitches. The winning quilt belongs to the fair. Other blocks are put into the quilts that are raffled off."

"I like to compete," says Laurie Rasmussen, who through the years has won three hundred and fifteen ribbons and seventeen rosettes for baking and floriculture. "No half effort for me. I got the bug in 1974. Even as a kid, I thought the carnival part of the fair was okay, but I liked the barns and exhibitions. I know what the judges are looking for: perfection. So I bake several pies or cakes, and then I choose the best.

"We're real rural here, and I like it that way, except that the U.P. gets neglected. It was even left blank on a seed catalogue zone chart, as though we just don't exist! And we've been left out of some children's history books. We call ourselves Yoopers, as in 'U.P.' Those in lower Michigan we call Trolls, 'cause they live under the bridge.

"My mother's parents owned the first telephone exchange around here, the Cooks Telephone Exchange. My great-grandfather started it in 1929. My mom was one of the operators. They had an old-time switchboard like you see on *Mayberry*, and cranks on the wall. My grandfather put up poles and strung lines, and I'd go with him to deliver the phone books. I guess the company served about a hundred families. When my mother and father were going together, they had a secret ring to get each other; two shorts. As it was a party line, of course people started listening in! The exchange went on till 1961, when the family sold it to General Telephone."

―――――――――

LAURIE RASMUSSEN'S ORANGE CHIFFON CAKE

"I've won several Grand Champion awards at the Upper Peninsula State Fair with this cake," says Laurie. "The cake becomes Lemon Chiffon Cake when you use grated lemon rind instead of orange rind."

MAKES 12 SERVINGS

2 cups sifted all-purpose flour	8 eggs, separated
1½ cups sugar	¼ cup cold water
1 tablespoon baking powder	2 tablespoons grated orange rind
1 teaspoon salt	½ teaspoon cream of tartar
½ cup vegetable oil	

- Preheat the oven to 325 degrees.
1. Sift together the flour, sugar, baking powder and salt into a big bowl. Make a well in the center. Pour in the oil, egg yolks, water and orange rind. Beat at low speed with an electric mixer for 1 minute, or until smooth.
2. In another bowl, and with clean beaters, beat the egg whites with the cream of tartar at high speed until stiff peaks form.
3. Transfer a scoop of egg white to the flour mixture, and mix it in with a rubber spatula to lighten the batter. Fold in the rest of the egg whites, until just blended.
4. Pour the batter into an ungreased 10-inch tube pan. Pull a table knife once through the batter to break up any large air bubbles. Bake in the center of the oven for 65 minutes, or until the top springs back when lightly touched. Invert the tube pan over the neck of a bottle to cool. When completely cool, turn the pan upright and remove the cake.

OVERNIGHT OMELET

"This is terrific for a Sunday brunch," says Laurie Rasmussen. "You can become creative and add other ingredients."

SERVES 12

2 dozen eggs	1 (4-ounce) can mushrooms, drained
½ cup milk	and chopped
1 teaspoon salt	1 pound cheddar cheese, shredded
1 (10¾-ounce) can cream of	½ cup chopped onion
mushroom soup	½ cup chopped green pepper

1. Beat the eggs, milk and salt together. Place in a skillet, preferably nonstick, and cook only until heated through and beginning to thicken, but still pourable. Don't scramble! Pour the mixture into a greased 9 × 13 × 2-inch ovenproof pan.
2. Mix together the remaining ingredients. Spoon evenly over the egg mixture and refrigerate overnight.
- Preheat the oven to 250 degrees.
3. Bake the omelet for 1 hour, or until the mixture is firm. Cut into squares to serve.

SHEBOYGAN COUNTY FAIR

Plymouth, Wisconsin

FAIR DATE: LABOR DAY WEEKEND, FOR 5 DAYS

GAZETTEER: *Take Highway 23 to Fairview Drive for the Sheboygan County Fair Park in Plymouth.*

Old view of the fairgrounds.

"People come early to the fair and they stay all day," says Lois Schreiber, fair manager. "And the exhibitors have to stay too; the animals have become so valuable you have to stay to watch over them. So our ecumenical church service on Sunday morning in the grandstand gives everyone a nice way to start Sunday. We work hard to make the service nondenominational. It's not unusual to have a couple of thousand people.

"We're big square dancers around here. You'll find more than four hundred dancing, all dressed up, with the ladies' skirts matching the men's shirts. We provide the space, the music and the caller, and the people come to meet their friends. After that, everyone's ready for beer and brats: bratwurst boiled in beer, then grilled and served on a bun. Wisconsin soul food!

"All the big-name acts that come through here specify in the contract that they want bottled water," says Ethel Holbrook, who cooks for the talent and their crews. "And they're looking for good home-style meals that won't set their systems off—they can't take that chance.

"In addition to special requests, we always supply a full meal. Here are a few of things entertainers have asked for: Barbara Mandrell asked that breakfast be served: Danish pastries, cereal, toast, assorted jams and jellies, and juice. Crystal Gayle wanted two types of steamed vegetables with dinner, hot tea and hot water with lemon. Johnny Cash specified prime rib with the end cut burnt. STYX, a rock group, requested linen tablecloths and napkins, china or stoneware service, and silverware. The Oak Ridge Boys asked for lettuce salad with red wine vinegar dressing.

"During both their evening performances, the Oak Ridge Boys told the audience of seven thousand how great the food was, and how they appreciated the wonderful service! In fact, all of the entertainers have been very appreciative. I met my husband at the Sheboygan County Fair," says Ethel. "My 4-H club was having a function, so I said to this boy, whose parents were friends of my mother and dad, 'Would you like to go?' He said, 'I'll give you a penny and you can send the invitation on a postcard.' A penny was what a card cost then. That was almost fifty years ago. We've had five lovely children and a happy marriage."

ETHEL HOLBROOK'S HAM LOAF

"It's really my aunt's recipe," says Ethel.

SERVES 6 TO 8

1	egg	1	tablespoon salt, or to taste
1½	cups milk		Freshly ground pepper to taste
3	slices white bread	½	cup packed brown sugar
1	pound ground beef	1	teaspoon dry mustard
¾	pound ground ham	½	cup pineapple juice
¾	pound ground pork		

• Preheat the oven to 375 degrees.

1. Beat the egg and milk together; soak the bread in the mixture for 10 minutes or longer.

2. Combine the meats with the soaked bread, salt and pepper, mixing them well but gently so as not to compress the meat. Put into a greased 9 × 5 × 3-inch loaf pan.

3. Make the topping. Mix the brown sugar, mustard and pineapple juice together. Spread this paste over the loaf. Bake 1 ½ to 2 hours, until the top is brown and bubbling.

The mayor milks a cow, 1949.

SHAWANO COUNTY FAIR

Shawano, Wisconsin

FAIR DATE: LABOR DAY WEEKEND, FOR 6 DAYS

GAZETTEER: *For the fairgrounds, take State Highway 29, which becomes Green Bay Street.*

From the premium book, 1932.

"I've eaten cheese every day of my life," says Charles O'Brian, secretary of the Shawano Agricultural Society and fair manager. "This is cheese country; the fair has cheese and butter judging, and cut cheese is given away. Our fair is well over a hundred years old. The Agricultural Society has run them all, through two world wars and the polio epidemic that closed the schools in 1958.

"The cows around here are mostly Holsteins. Eighty percent of the people here are from Pomerania, the Holstein area of Germany. They settled here because of the similarity; rolling hills and wooded forests. This country took a long time to clear. In the past ten years the fair has honored two hundred and ten of what we call 'century farms'—farms that have been in the same family for a hundred years.

"Before milking machines, the average herd was around ten cows; now it's around fifty. We have a couple of cheese factories here that pick up the milk every day, combine it and make cheddar. The cheesemakers can tell the season, and what the farmers are feeding, by the taste of the milk. They know immediately when cows that have been in the barn all winter eating silage go out in the spring pasture. And more—they'll say, 'Louis must have opened up a new silo today, because I can smell and taste the difference in the milk.' Basically, none of that affects the cheese, which is heated, fermented and aged.

"We have three entertainment areas at the fair. One is strictly for polkas, both German and Polish, alternating with country western. Rainbow

Toy tractor pull.

Valley Amusements runs our carnival. They're a Wisconsin outfit. Our state is noted for circuses and carnivals; some two hundred got started here, including Ringling Brothers. Carnivals are the offspring of circuses—trapeze artists and other circus acts started performing at fairs.

"In our Tractor Pulls we use only local farm tractors. The pullers have to be Shawano County residents, and we allow no soup-ups.

"We have thirty-five thousand humans in the county, including nine cheese-makers. And we have probably close to fifty thousand cows."

Henrietta Peters, a retired dairy farmer, originated the Dairy Foods Promotion Class at the Shawano County Fair. "The local milk company puts up money for prizes," she says, "and the cheese is also a prize!"

BONNIE ENGEL'S BACON-CHEESE MUFFINS

Bonnie won a red ribbon in 1992 with these muffins. You may wish to reduce the sugar to 1 tablespoon.

MAKES 12 MUFFINS

½ pound bacon
  Vegetable oil
1 egg, lightly beaten
¾ cup milk
1¾ cups all-purpose flour

¼ cup sugar
1 tablespoon baking powder
1 cup shredded Wisconsin cheddar cheese
½ cup Grape Nuts cereal

- Preheat the oven to 400 degrees.
1. Cook the bacon until crisp. Drain on paper towels. Measure the drippings; if needed, add enough oil to make 1/3 cup. Combine the drippings with the egg and milk and set aside. Crumble the bacon and set aside.
2. In a large mixing bowl, stir together the flour, sugar and baking powder. Make a well in the center and add the dripping-egg mixture all at once. Stir just until moistened; the batter should be lumpy. Lightly fold in the crumbled bacon, cheese and cereal.
3. Fill greased or paper-lined muffin cups about 3/4 full. Bake for 15 to 20 minutes, or until lightly browned.

WISCONSIN CHEESE FUDGE

Cheese fudge is a popular category at Wisconsin fairs. This version won Marion Voelz a prize at the Shawano Fair.

MAKES 1 PAN OF FUDGE

2 cups shredded Wisconsin cheddar cheese, or other cheddar

½ cup cocoa powder

1½ cups nonfat dry milk powder

½ pound (2 sticks) butter, at room temperature

1½ pounds confectioners' sugar

1 teaspoon vanilla extract

¼ cup whole milk, if needed

1. Have all ingredients at room temperature. Combine the ingredients in the large bowl of a standing electric mixer and beat until creamy. (It may be necessary to add some or all of the whole milk.) The fudge can also be made in a food processor equipped with the metal blade.

2. Butter a 9-inch-square baking pan. Press in the fudge and chill for several hours or longer. When firm, cut into 64 squares. The fudge should be kept frozen, or very cold. For a cake frosting, simply thin it with more milk.

HEART OF ILLINOIS FAIR

Peoria, Illinois

FAIR DATE: MID JULY, 9 DAYS

GAZETTEER: *Peoria is right in the center of the state, in an area of agriculture and light industry, including the Caterpillar Tractor plant. Take University Street, off I-74, to Northmoor.*

"That old expression 'Does it play in Peoria?' came from vaudeville's heyday," says Eileen M. Frye, the fair's general manager. "It meant if a show went over here, every other town would like it too. As for our fair, it really *does* play in Peoria!

"We're proud to have held the line on admission—one price for everything, including top entertainment; you only pay for food and rides. Of course, sometimes the entertainment surprises us all, like the time a parachuter landed on a pig. Hard on the dignity, but both were okay.

"Group effort is a big part of our fair. In Operation Earthmover, in 1948, using donated equipment, the Corps of Engineers moved more than half a million cubic yards of earth in a twenty-four-hour period to make the base of our fairground.

"Lots of high school alumni come for the FFA tug-of-war—it's especially important to the rural high schools. We have about eight teams—schools against schools, and alumni against alumni."

"I grew up on a farm in Peoria County," says Virginia Williams. "My mother, in the Peoria Garden Club, was active years ago when the fair was in tents, before the buildings went up. I got involved, too; I love to cook and to bake, and my entries just snowballed. One year I had twenty-four. I have a husband and four children, and before fair time they knew I'd be doing nothing but baking. I'd freeze the best. 'Pick an ugly one,' I'd say about a cake, when the hungry kids walked in—I said it so much that 'Pick an ugly one' is an all-purpose joke in our family. One year I decorated a winning cookie jar and presented it to Pat Boone. A few years ago the Peoria Pork Producers sponsored a cookoff. You had to prepare your entry right at the fairground. Besides taste, you're judged on preparation and place set-

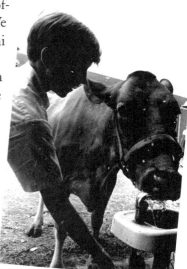

A cool drink.

ting. We had to serve the three judges. I won two years in a row with my marinated pork tenderloin. This is Illinois, and pork—'the other white meat'—is special."

VIRGINIA WILLIAMS' MANDARIN PORK EMPRESS

Virginia serves slices of tenderloin on a bed of wild rice, spoons on the
Mandarin Orange Sauce, and garnishes each plate with a sliced orange twist, a kiwi slice
and a whole strawberry.

SERVES 6 TO 8

THE MARINADE:

½ cup soy sauce

¼ cup honey

2 tablespoons fresh lemon juice

½ teaspoon powdered ginger

1 cup vegetable oil

½ cup juice drained from can of
      mandarin oranges

3 whole pork tenderloins, each about
      8 × 2½ inches, about 1 pound

THE SAUCE:

1 cup orange juice

½ cup packed brown sugar

½ teaspoon ground ginger

1 tablespoon cornstarch mixed
      with 2 tablespoons water

2 tablespoons butter

1 can mandarin oranges, drained

Virginia Williams, with Jennifer and Amy,
filling a cookie jar for Jim Nabors, 1979.

1. Whisk together all the marinade ingredients. In a nonaluminum container, let the tenderloins marinate in the refrigerator for several hours or overnight.

2. Drain the meat. In a covered-kettle barbecue grill, cook over moderately hot coals for approximately 1 hour, or until the meat reaches 170 degrees on an instant-read thermometer. Baste frequently with marinade.

3. Make the sauce: Combine the orange juice, brown sugar and ginger, and cook until bubbling. Stir in the cornstarch mixture and cook, stirring, until the sauce thickens. Off the heat, add the butter and the orange slices. Cut the meat into 1-inch slices, and spoon on the sauce.

THE SANDWICH FAIR

Sandwich, Illinois

FAIR DATE: SEPTEMBER, WEDNESDAY AFTER LABOR DAY, FOR 5 DAYS

GAZETTEER: *The fairground is west of Sandwich city limits, on Suydam Road, sixty miles southwest of Chicago.*

"This is prime Illinois agricultural land," says Louis Brady, fair secretary, who has been with the fair since 1939. "Dekalb Genetics produces hybrid seed corn here. We produce barbed wire, too; barbed wire started here in Dekalb County. It was the hub of manufacturing, and different patents were held by many.

"We're an old-fashioned agricultural fair, but we move with the times. We have bungee jumping, and llamas in the livestock events, and photography competitions. But we brought back some of the good things from the past—horseshoe pitching on six courts, and draft horses."

Philatelists the world over send requests for stamp cancellation to the Sandwich Fair, which has its own post office on the fairground. Artist Quen Carpenter

Draft horse show, 1948.

designs the stamp with a different theme for each year. In a recent year she chose the melodramas, old-time "cheer the hero, hiss the villain" shows and vaudeville acts presented nightly by the Indian Valley Theater, featuring such classics as *Nellie Was a Baker, 'Cause She Kneaded the Dough.*

KRINGLA COOKIES

Ruth Marselus says, "I've been going to the fair since I was a little girl—we were a farm family of eight, so we always went when there was one price, on Family Day, a Thursday. I just love baking, and I'm never without these cookies I learned from my Norwegian mother. I've won blue ribbons with them, too!" We have added the almonds and the nutmeg.

MAKES ABOUT 40

8 tablespoons (1 stick) butter, at room temperature

½ cup packed brown sugar

½ cup granulated sugar

2 teaspoons vanilla extract

1½ teaspoons grated lemon peel

1 large egg

3 cups all-purpose flour

1 cup finely ground almonds

1 teaspoon baking soda

2 teaspoons baking powder

2 teaspoons freshly grated nutmeg

1 cup sour cream

1. Using a standing electric mixer on low and then moderate speed, cream the butter and sugar together until light and fluffy. Beat in the vanilla, lemon peel and the egg.
2. Whisk together the flour, almonds, baking soda, baking powder, salt and nutmeg. With the mixer on low speed, alternate adding the flour mixture and the sour cream, beginning and ending with flour. Wrap the dough in plastic wrap and chill overnight.
- Preheat the oven to 350 degrees.
3. Work with about 3 tablespoons of dough at a time, and keep the rest chilled. On a lightly floured surface, roll out the dough. Cut off strips about 3¹/₂ inches long and 1 inch wide, and roll them into pencil shapes. Place them on a greased baking sheet, forming the pencils into figure eights.
4. Bake about 8 minutes, or until lightly browned; don't let the Kringlas get too dark or they will be dry.

SIX

THE PLAINS

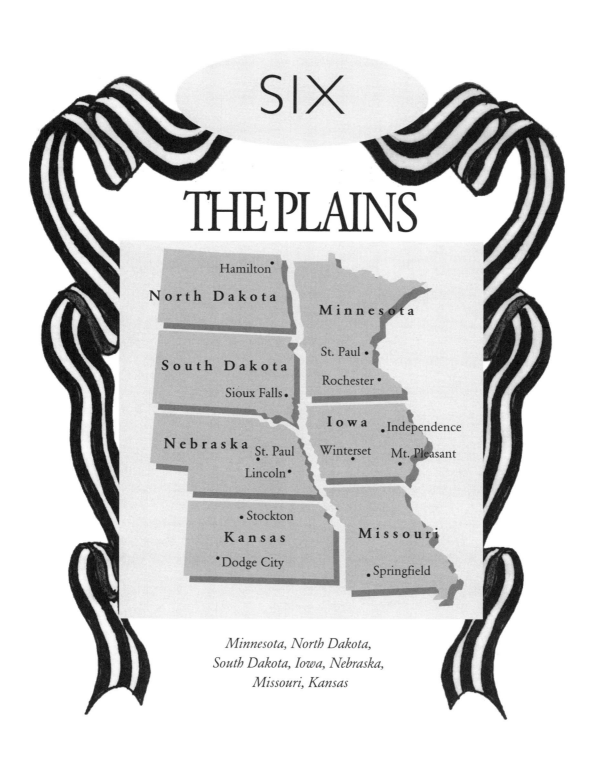

Minnesota, North Dakota,
South Dakota, Iowa, Nebraska,
Missouri, Kansas

MINNESOTA STATE FAIR

St. Paul, Minnesota

FAIR DATE: LATE AUGUST, 12 DAYS, THROUGH LABOR DAY

GAZETTEER: *The 310-acre fairgrounds are located at 1265 Snelling Avenue N, St. Paul.*

Radio fans of Garrison Keillor's *A Prairie Home Companion* have fond memories of the Minnesota minnesingers' broadcasts from the fair. For Keillor, as for thousands of other small-town kids, it was the high point of summer, before school and routine again took over their young lives.

First held in 1859, and moved to the site of a defunct poor farm in 1885, the fair is a Rite of Fall for all of Minnesota. It's the chance to indulge in the fair favorite, deep-fried cheese curds (don't say a word till you've tried them), and giant Bavarian pretzels with garlic butter and cheese. To work off some of the cholesterol, you can bungee jump.

Or you can just stroll through the fairgrounds, taking in the sights and exhibits, such as the indigenous fish, plants and wildlife at the Minnesota Department of Natural Resources display in a log cabin erected by the WPA. "Back in the Thirties," says media director Jerry Hammer, "the head of the WPA in Minnesota was a great fan of the fair, so most of the exhibit halls were built then, as a way to keep people employed. They did a good job, too. We have streets, sidewalks, curbs and gutters, which you don't normally see at fairs, so bad weather doesn't have that much of an impact."

The fair has all the agricultural competitions and big-name entertainment anyone could wish, and it also honors the crafts. Displayed at the Crop Art Show are paintings fashioned from tiny seeds and bits of flowering plants. There is artwork sculpted from beeswax, too, in the Bee and Honey area, and ethnic handicraft demonstrations, such as the Norwegian embroidery called Hardanger work.

For sculpture in another medium, the fair has its own Butterheads, sponsored by the American Dairy Association of Minnesota. "Various regions of the state send their dairy princesses as candidates for 'Princess Kay of the Milky Way,'" says Jerry Hammer. "Each day of the fair, one of these kids puts on a ski jacket and climbs into a glass-sided cooler with an artist

Linda Christensen, sculptor, begins a Butterhead. (The temperature is 38 degrees F.)

who sculpts her head out of an eighty-five-pound block of butter. Eleven girls in all sit for their portraits, seven hours at a time, including the one chosen as Princess Kay. The case is called the Glass Gazebo, and people can watch the sculptor at work. Once the fair is over, the princesses take their heads home and donate them, or whatever."

PICKLE POWER, MINNESOTA STYLE

In 1990 Gedney, the Minnesota pickle packers (since 1881, the labels proudly state), had a bright idea. Why not take the blue ribbon pickle winners at the state fair, reproduce the recipes, and send them forth to a fine commercial future? So they did. Were Genevieve Spano, Nita Schemmel and Maxine Wagner jarred by success? Tickled pink is more like it. They're good marketers, too. Maxine Wagner sometimes concludes a telephone conversation with a hearty "Buy more pickles!" Nita asks people in the supermarket, "By the way, have you tasted my pickles yet?"

The Minnesota ladies are not only great pickle makers, they're *clean* picklers. Maxine buys the smallest cukes she can find at the Farmers' Market and scrubs each one with a toothbrush (reserved especially for this task). Nita uses a washcloth to wipe her cucumbers. Genevieve, however, leans toward mechanization; she cleanses her cukes in the washing machine. "Use cold water and throw in half a dozen hand towels," she advises. "Add half a bushel of small to medium cukes, run the full cycle, but *do not let them spin*. Drain, take out the towels, and put the cukes through a rinse cycle; again, no spinning. In half an hour you have all those pickles washed." Putting up 150 quarts of dills is nothing to Genevieve, who believes in lots of pickles.

Winners on pickle labels.

NITA SCHEMMEL'S
BREAD AND BUTTER PICKLES, PLAIN OR HOT

"The recipe for plain bread and butters was given to me at a bridal shower," says Nita.
"Adding the hot peppers was my idea. My husband Jim plants the cukes each spring."

MAKES 4 PINTS

THE PICKLE MIX:

12 medium cucumbers, about 7 inches, sliced thin

12 medium onions, sliced thin

½ green bell pepper, seeded and sliced thin

¼ red bell pepper, seeded and sliced thin

Scant ½ cup pickling salt

8 small Santa Ana hot red peppers, sliced thin (optional)

THE SYRUP:

2 cups cider vinegar

2 cups sugar

1 teaspoon celery seed

1 teaspoon mustard seed

¼ teaspoon turmeric

1. Prepare the pickle mix: Combine the cucumbers, onions, green and red bell peppers, and salt (including the hot peppers, if you wish) in a large nonaluminum bowl with two trays of ice cubes. Let soak, unrefrigerated, for 2½ hours.

2. Make the syrup: In a large nonaluminum kettle, combine the vinegar, sugar, celery seed, mustard seed and turmeric and heat just to boiling. Drain the pickle mix and add to the boiling syrup. Lower the heat and simmer, uncovered, until the cucumber slices are yellow and translucent, about 25 minutes or longer, stirring often. Seal in sterilized jars and process in a water bath, according to manufacturer's instructions, for 5 minutes.

AND THE WINNER IS...

"My business is the national and regional promotion of consumer products," says media spokesperson Joyce Agnew, president of Agnew Communications, Inc., "and to my knowledge, I'm the only person specializing in the management of recipe contests at state fairs.

"One of my clients is SPAM, which has contests at almost sixty fairs. We've had people enter SPAM fettuccine, SPAM cheesecake, SPAM mousse. The competition is serious. The winner gets one hundred dollars, and all entrants get ribbons or award certificates. The SPAM collectibles, though, like aprons and mugs, are more valuable to these people than the money. It's prestige.

"When the corporations who sponsor the contest get the winning recipes back to their test kitchens, they get a good view of what consumers are doing with their products, and they can spot national trends. For example, one year we saw a lot of apricots in quick breads, and the next year it was poppy seeds.

"I've talked to ladies who have over two thousand ribbons; they bring their entries to the fair in station wagons. Then you'll come across a young woman who enters for the first time, and wins. There are lots of family entrants of mother, daughter and grandmother.

Joyce Agnew.

"I will never figure out what makes some people able to win all the time, like Marjorie Johnson in Minnesota. I've done these programs for years, and the same names keep popping up. I ask, 'What is your secret?' and the answer is 'I don't know!'"

DORM LIFE AT THE 4-H BUILDING

Meet Chester, Lester, Hildegard and Elaine, a few of the workers in the 4-H building that feeds seven hundred kids at a time during the run of the Minnesota State Fair. These workers are getting on—most of them are over fifty years old—and they aren't getting paid, either. That's because each one is a piece of equipment, mostly put in when the Works Progress Administration built the structure in 1939. Chester and Lester (known as The Boys) are massive ovens. Hildegard the Dishwasher dates to the early part of the century; she may have come from an old CCC camp. She can still heat the water hot enough to pass health inspection, however. O'Riley the Refrigerator, Elaine and Gus, the walk-in coolers, and Jumbo the Freezer keep chugging along, too. Sue Boehland, manager of this, the largest kitchen on the fairgrounds, says, "We speak softly to them so they'll keep on running."

The building was refurbished in 1989, on its fiftieth anniversary. There's a huge exhibit hall on the first floor; the kitchen, dining room and auditorium are on the second; and on the third floor, eighty-eight tall steps above the first floor, are the dorms, known as the "4-H Hilton." They hold nine hundred bunks, half for boys, half for girls. The dorms are home to 4-H'ers who come in groups from their counties, stay for several days, exhibit their projects, and make new friends and great memories.

MARJORIE JOHNSON'S APRICOT ALMOND BRICKLE COFFEE CAKE

"My coffee cake won the Land O'Lakes Light Sour Cream Quickbread contest at the Minnesota State Fair in 1991," says Marjorie. "I have many ribbons, but each one is exciting, even though I have over one thousand now. Minnesota is full of accomplished bakers, so it is a challenge each year to bake my best.

"I spend the summer baking. My husband knows he has to do the vacuuming! I first entered the state fair in 1974, and our local Anoka County fair in 1986. The county fair comes about three weeks before the state fair, so I sometimes use it as a tryout for my state fair entries. I usually take about fifty baked entries to each fair."

MAKES 2 COFFEE CAKES

THE CAKE:

12 tablespoons (1½ sticks) butter, at room temperature	1½ teaspoons baking powder
1½ cups granulated sugar	1½ teaspoons baking soda
4 large eggs	¼ teaspoon salt
1 teaspoon vanilla extract	1½ cups light sour cream
1 teaspoon almond extract	½ cup dried apricots, chopped
3 cups all-purpose flour	½ cup toasted chopped almonds
	½ cup almond brickle chips

THE GLAZE:

½ cup confectioners' sugar

¼ teaspoon almond extract

2 to 3 teaspoons milk

Snack time.

- Preheat the oven to 350 degrees.
1. Grease and lightly flour two 9-inch layer cake pans. Set aside.
2. Using an electric mixer, cream together the butter and the sugar until light and fluffy. Add the eggs, one at a time, and beat well after each addition. Beat in the vanilla and almond extract.
3. Sift the flour, baking powder, baking soda and salt together. Stir the flour mixture, alternating with the sour cream, into the butter mixture, beginning and ending with the flour.
4. Put one-quarter of the batter in each pan. Sprinkle half the apricots over each batter, followed by one-quarter of the nuts and one-quarter of the brickle chips. Divide the remaining batter and pour it into the pans, followed by the remaining nuts and chips. Bake for 35 to 40 minutes.
5. Meanwhile, prepare the glaze by combining the sugar and almond extract in a small bowl and adding just enough milk to make a spreadable consistency.
6. Cool the layers on racks. Drizzle the cooled cakes with the glaze.

MIDWEST SPAM SALAD

Rosann DeVoe entered her dish in the Stephenson County Fair SPAM contest in Illinois.

MAKES 12 SERVINGS

1 cup salad dressing (such as Miracle Whip)	1 cup diced mild cheddar cheese
1 tablespoon prepared mustard	½ cup chopped bread and butter pickles
3 tablespoons sugar	½ cup finely chopped onion
¼ teaspoon pepper	3 hard-boiled eggs, chopped
2½ pounds frozen green peas, thawed and drained	1 (12-ounce) can low-salt SPAM, diced
	Lettuce leaves, if desired

- In a large bowl, combine the dressing, mustard, sugar and pepper. Add the peas, cheese, pickles and onion. Toss gently to mix. Add the eggs and SPAM and stir to combine. Cover and refrigerate until chilled, at least 2 hours. Serve on lettuce leaves if desired.

OLMSTED COUNTY FAIR

Rochester, Minnesota

FAIR DATE: LATE JULY, EARLY AUGUST, 7 DAYS

GAZETTEER: *From Minneapolis, take Highway 52 to Rochester, home of the Mayo Clinic. Then go east on 14, and at the junction of 63 and 14, you can see the fairgrounds.*

The spark plug of the present Olmsted County Fair was George F. Howard, County Superintendent of Schools early in the century. "When visiting my schools in the spring of 1904," he wrote, "I carried a sack of seed corn in my buggy. I enrolled boys to plant 200 grains of corn, and make an exhibit of the crop in the fall. The following year I added bread-making and cake-baking as a new work for the girls." The exhibits were held in the German Library Hall in Rochester until the fall of 1907, when businessmen urged Howard to move his county school fair to the old fairgrounds. "The county fair had been dead for many years, run into the ground by horse-racing and bad management," said Howard. "I organized a County Agricultural Society to run the fair, rented the old fairgrounds for a week and opened the Olmsted County School Fair. The fair was a success."

The fair continues to be a success, and young people still exhibit, now under the aegis of 4-H. Older folks bang culinary utensils in the Senior Citizens' Kitchen Band and raise their voices with the Young at Heart singers. "We have square dancing, and entertainment every night on the Flatbed Stage," says Jerry Hennessey, fair secretary. "We're a 'wheel fair,' with stock car races, Demolition Derby, and a monster truck show. But we also have one of the largest draft horse shows in the nation."

Olmsted fairgoers have always had a taste for engine power. According to the fair's historical record, in 1911 "An array of the latest automobiles attracted almost as much attention as the prize-winning cattle and swine." When famed acrobatic aviatrix Ruth Law came to barnstorm over the fair, "The cost was $300 per day for four days." That was a hefty sum, and Miss Law insisted upon being paid at the end of each day, cashing the check the following morning. One the fourth day, planning to leave town after the bank closed, she asked for cash. "Mr. Cutting [Secretary of the Fair Board] brought her $300, nearly all in one dollar bills. She accepted it very graciously."

Dick Billings' six-mule hitch attracted much attention in 1960.

A FESTIVE OLD-COUNTRY TREAT

Irish, Germans and Norwegians are a strong presence in Olmsted County. Ida Thronson's parents were both born in Norway. Mrs. Thronson was a 4-H leader for many years, and her blue ribbon specialty, often exhibited, is Hardanger work, a type of fancy Norwegian embroidery, which she teaches. "It has to be perfect, and I enjoy every minute," she says. "I've made altar, baptismal and pulpit cloths for the Lutheran church. And when I take my Rømmegrøt down to potlucks at the senior center there, I never come home with any."

IDA THRONSON'S RØMMEGRØT (CREAMY SWEET PUDDING)

"My Norwegian mother always made Rømmegrøt with raw unpasteurized cream, just as they did in the old country. You can use heavy cream if you like, but I find half-and-half rich enough! A little of this dessert goes a long way. Rømmegrøt is for festive occasions, like Christmas and the 17th of May, Norwegian Independence Day."

SERVES 4 TO 6

8 tablespoons (1 stick) butter
½ cup flour
   Dash of salt

2 cups half-and-half
   Sugar to taste
   Cinnamon to taste

1. In a large saucepan, preferably nonstick, melt the butter over moderately low heat. Stir in the flour and mix until smooth; continue stirring for 2 minutes. Add the salt and slowly add the half-and-half. Whisk until smooth.
2. Raise the heat and let the mixture come to the boil. Let it boil for about 5 minutes, whisking frequently until the mixture thickens and the floury taste is gone. Serve warm, sprinkled with sugar and cinnamon.

PEMBINA COUNTY FAIR

Hamilton, North Dakota

FAIR DATE: THIRD WEEK IN JULY FOR 5 DAYS, WEDNESDAY THROUGH SUNDAY

GAZETTEER: *Hamilton is halfway between Winnepeg, Canada, and Grand Forks. The fairgrounds are on old Route 81, just off I-29.*

"We're the oldest continuous fair in North Dakota," says Fern Peterson, who was chairman of the centennial fair project in 1993. "We've been out of money, but we've always been able to have a fair.

"We're very rural around here, and people really appreciate horse racing. They come from miles for that. We have harness racing and also chariot racing. The chariot is usually a barrel mounted on two wheels pulled by two horses. When the horn blows, do they go around that track! It's all local people, who race for grins and kicks. Some even get themselves up in Roman costume.

"For the centennial pageant, we went back in history to redo major events in song and narration. We always had clowns, so we had them in the stands. Older folks, got up in overalls, sang songs: 'Sentimental Journey,' and 'Oh Dear, What Can the Matter Be?' We used to have girlie shows on the midway, a tent where the ladies came out, wiggled and waggled around and cried, 'Hurry up! The show's about to start!' and the men of course lined up. We did a little feature on that. The finale was the Centennial Queen, Dawn Michelle Hurst, who came down the track as a vision of the next century."

Minnie Briese, who with her husband is retired from farming and the restaurant business, represents the Blue Ribbon Homemakers at the fair. "We've been together over forty-five years," she says. "We learned how to make pies and bread, and when I first started, they showed us how to sew, and even make hats! So many are getting up in years now. I love the fair, and I've won a lot of ribbons and silver plates. My husband Marvin is a jack of all trades—I used to help him out in the fields. Now I'm president of the Senior Citizens, and president of the Ladies' Aid Society at the Lutheran Church."

MINNIE BRIESE'S POTATO SOUP

"I was noted for my soups when Marvin and I ran The German Cafe in Hamilton, North Dakota," says Minnie Briese. "This soup was the favorite, and I made a fresh kettle every day." The recipe makes a lot of soup, but you can freeze some in batches.

MAKES ABOUT 5 QUARTS

3 cups mashed potatoes, leftover or freshly made
4 quarts water
2 large raw potatoes, peeled and diced
3 large carrots, shredded
4 stalks celery, diced
1 large onion, chopped
4 whole allspice berries, crushed
1 large bay leaf
Salt and pepper to taste
1 medium onion, diced
8 tablespoons (1 stick) butter
8 slices bacon, diced
1 cup sour cream
½ cup flour
2 cups cream

1. In a large heavy kettle, combine the mashed potatoes with the water. Add the raw potatoes, carrots, celery, large onion, allspice, bay leaf and salt and pepper. Bring to the boil and cook uncovered until the vegetables are tender.
2. In a large skillet, cook the medium onion in the butter until the onion pieces are clear. Add the bacon and fry until brown. Pour the contents of the skillet into the soup kettle.
3. Mix the sour cream, flour and cream together, leaving the mixture slightly lumpy. Add to the soup and simmer for five minutes.

Marvin and Minnie Briese.

MINNIE'S ROULADEN
(BRAISED STUFFED BEEF ROLLS)

"We featured this German dish at our restaurant," says Minnie. "Very good!
Serve it with mashed potatoes."

SERVES 4

4 (7-ounce) slices round steak, pounded
    ¼ inch thick (each slice about
    4 × 8 inches)
Salt
Freshly ground black pepper
4 teaspoons prepared dark mustard
2 slices bacon, cut in half crosswise
1½ cups chopped onion
1 sour dill pickle, cut lengthwise into 4 pieces

1 tablespoon bacon dripping
1 (10½-ounce) can beef consommé
10½ ounces water
1 bay leaf
1 tablespoon flour
1 tablespoon water
2 teaspoons butter, at room
    temperature
Parsley, for garnish

1. Lay out the slices of beef on a work surface and season them lightly with salt and pepper. Spread a teaspoon of the mustard over each slice, and place a half-slice of bacon down the center. Add 1 teaspoon of the chopped onion and a pickle slice to each piece of meat and roll into a tight cylinder, jelly roll fashion. Secure the ends with kitchen cord, and fasten the side closed with a small metal skewer or a round toothpick.

2. Heat the bacon drippings in a heavy pan over moderate heat. Saute the rouladen, turning frequently, until brown all over. Add the remaining chopped onion and cook, stirring, for three minutes. Add the consommé, water and bay leaf, and bring to the boil, stirring in any browned bits clinging to the pan. Lower the heat, partially cover, and simmer the rouladen for 1 hour and 40 minutes, or until the meat is tender when pierced with the tip of a knife. Turn the rouladen a few times as they cook.

3. Work the flour, water and butter together. Remove the rouladen and bay leaf from the pan and stir in the flour-water-butter mixture. Bring the gravy to the boil, lower the heat and simmer, stirring, until the gravy has thickened. (If it appears too thick, thin with a little water.) Clip the kitchen cord off the rouladen and return them to the gravy to warm through. Serve garnished with sprigs of parsley.

SIOUX EMPIRE FAIR

Sioux Falls, South Dakota

FAIR DATE: SECOND WEEK OF AUGUST, FOR 6 DAYS

GAZETTEER: *Southeastern South Dakota is a region of diversified agriculture, divided from the ranching, western of the state by the Missouri River. The fairgrounds are near the intersection of I-29 and I-90, just west of Sioux Falls.*

"Ours is now a regional fair, but it still qualifies as the Minnehaha County fair," says Shirley Alberty, in charge of domestic and fine arts at the fair. "My husband is in charge of the Old Mac-Donald's Farm, for the children. We met on a 4-H trip to Chicago, some forty-five plus years ago. All five of our kids were in 4-H.

"One day of the fair is called Farmers Appreciation Day, organized by the Chamber of Commerce. It's a way to honor the farmers, and thank them for what they contribute to our lives. We give lunch to about five thousand farm people. It's a way of showing appreciation."

"I'm just an old farmer's wife,' says Luella Larson. "A farmer's wife does a lot of cooking and baking. I also helped my husband in the field; drove a tractor, and enjoyed it. Years ago there was more work for the farmer's wife. I took care of the chickens, and I fed the hired man who lived with us and was treated as a member of the family.

"At harvest we furnished meals to the extra help, probably twelve to fifteen. We got up at five a.m. and had a big breakfast. We had forenoon lunch at nine-thirty out in the field, and then the big meal, dinner, with meat, potatoes, vegetables and pie, at home in the middle of the day. We also took our afternoon lunch out in the field in a pail, and ate it about three-thirty—a sandwich, cake, and lots of coffee, like the forenoon lunch. Nothing tastes as good as lunch out in the field. Then there was supper. Nowadays, a lot of farmers' wives work away from the farm. Need the extra paycheck, I guess.

"We live on my husband Barney's grandfather's 1876 homestead. They first lived here in a sod hut. Barney likes to brag that he is sleeping in the same room he was born in.

"For the fair I bring in forty to fifty entries every year. I started about twenty years ago, and I was lucky enough to win the purple ribbon—the best—with spritz cookies made from my mother's recipe. I was so proud! I'm ashamed to say I even put that ribbon on top of my basket, hoping everyone would see it as I went to the car at the end of the fair.

"On the day the entries go in, I want no one around. My husband has to fend for himself. You pack it all up carefully, hope nothing happens on the way, and get the entries in by six p.m. Then I ask myself, 'Why do I do this?' But next year it's the same old thing!"

LUELLA LARSON'S SPRITZ COOKIES

"My background is Swedish," says Luella, "and so are these cookies."

MAKES ABOUT EIGHTY 2-INCH COOKIES

1 cup sugar

1 pound (4 sticks) butter (no substitutes),
　　at room temperature

1 large egg

¼ teaspoon salt

4 cups flour

1 teaspoon almond extract

- Position 2 racks in the oven. Preheat the oven to 350 degrees.
1. Preferably using a standing electric mixer, cream together the sugar and butter until the mixture is light and fluffy. At moderate speed, add the egg and the salt. Add the flour, a little at a time. When it is incorporated, beat in the almond extract.
2. Do *not* refrigerate the dough; you want it soft. Pack it into a cookie press and press in rings or other designs onto two ungreased baking sheets.
3. Place in the preheated oven and bake for 5 minutes, then switch the positions of the sheets. Bake for 10 to 20 minutes in all, until the cookies are golden and lightly browned on the edges. Cool on racks. Continue baking until all the dough is used.

MIDWEST
OLD THRESHERS REUNION

Mt. Pleasant, Iowa

FAIR DATE: 5 DAYS, ENDING LABOR DAY

GAZETTEER: *The Midwest Old Settlers and Threshers Association, a nonprofit museum, sponsors an annual reunion for old machinery buffs. Henry County "Where Hospitality Is Habit," is corn-, soybean- and hog-raising land in southeastern Iowa, 50 miles south of Iowa City. For information on points of interest, visit the Henry County Welcome Center on Highway 34, across from the courthouse. The Midwest Old Threshers is located at the intersection of U.S. Highways 218 and 34, at 1887 Threshers Road.*

"The Association began in 1950, with fifteen steam engines and eight separators," says Sam Wynn, director of public relations. "Now we have two Heritage Museums with scores of steam traction and stationary engines, antique tractors and farm implements. At the reunion, you'll find over a hundred traction engines and a thousand gas engines; all are in working order. There's antique cars and trucks, and a huge craft fair. And threshing demonstrations, which people love to see."

Traction engines are steam driven; gears turn the wheels. They were expensive; a wealthy person would buy one and go from farm to farm, charging a fee or a percentage of the crop to thresh. A threshing machine separates grain from straw. Bundles of wheat or oats are fed into cylinders armed with beaters. Teeth rake away the grain, which is then winnowed–blown against a sieve–to separate the wheat from the chaff. The chaff blows away, and the straw blows out of a long tube into a stack behind the thresher. The grain pours out of a spout, caught in wagons or in burlap bags by men at the ready. A huge belt runs from the traction engine to the threshing machine, and drives it. Case Corporation of Racine, Wisconsin, was the biggest manufacturer of traction engines. Wood Brothers made them too. Crews were made up of twelve to twenty men, each

A 110 HP Case steam tractor engine rumbles into life.

with his own task. One ran the engine, another kept an eye on the belt, others loaded or collected grain. It could take two or three days to thresh–or thrash–a crop. "Now that kind of threshing is probably only done in Amish villages," says Wynn. "Today huge combines do the job of ten men, and much faster."

"I grew up with threshing on our farm in Minnesota," says Joyce Agnew (page 203). "My dad loved steam engines; he bought a thresher and a tractor, and threshed all over the neighborhood. When I was a kid in the early Fifties, he had a crew of about twelve, and they probably all had farms of their own. I don't think he was paid, though earlier he probably was. Threshing was exciting, beautiful and very dangerous. We were warned never to go near the belt. I remember the chugging, and the pulsation; almost like a vibration in the ground that you could hear and even feel in the house.

"The farmers in the crew milked their cows, and had a big breakfast at home. My mother did the cooking for everyone. At nine-thirty the men came in for morning coffee; that also meant cakes, cookies, rolls, the works. We'd barely done the dishes when we'd start dinner, which was at noon. I recall being seated at our long dining room table. The tons of food–meat, potatoes, gravy, vegetables, pies–when I look back on it, I can't believe we ate what we did. And none of those men were fat. My family raised cows, so we had a freezerful of beef. Mom would fry two big pieces of steak for each man, fry them till there was nothing left, like they did back then, or make meatballs. Mashed potatoes, and lots of gravy. We had vegetables out of a can; creamed corn, beans and peas. There had to be fresh pies, and a good farm wife had to make the pies that day.

"Mom and I washed the dishes till it was time to bake a cake (it had to be fresh too) for afternoon coffee, at three-thirty. She used cake mixes, and her own scratch frosting, which was delicious. The meal was coffee and 'thrash cake,' fresh bread, butter, jam and meat sandwiches. A full meal, in fact.

'My mom went to work when I was twelve, so that I took over feeding the crew, which was down to five or six by then. Mom made long, very detailed lists of what to do–'at ten a.m. peel the potatoes. Get the meat from the freezer'–that sort of thing. No recipes. I was smart enough to follow directions, but not smart enough to tell her I wasn't a slave."

Iowa farmers Bob and Betty Carlson volunteer in the Pancake Tent at the Midwest Threshers Reunion, for the benefit of the Old American Legion Annex. "We fry about three thousand pancakes," says Betty, "and countless eggs. Our son Don has Belgian draft horses, so he demonstrates threshing the really old-fashioned way, before steam. Six teams are needed, to form what's called a 'threshing ring.'

"I remember, growing up, how they'd go through the oat and wheat fields with binders, to tie up the grain in bundles. They'd set the shocks upright, and collect them in a wagon, ready for the threshing.

"We would get cold water from the well and fill up brown jugs. We covered them with soaked gunnysacks, and the evaporation kept the water cool. The children took water to the men in the field.

"Several neighbor ladies came in to help my mother with the cooking. In the 1930s, we didn't have electricity in the house. Mother ordered a huge piece of beef for roasting from our small-town grocery, and a huge chunk of ice, to keep things cold and for the lemonade served along with coffee at mealtime. There was a mountain of mashed potatoes, and a sea of gravy, and lettuce and vegetables from the garden; corn, green beans, creamed peas. Mother made apple and other fruit pies.

"For the men to wash up before dinner, my father set up planks on sawhorses. He took laundry tubs and filled them with soft water from the cistern, that warmed in the sun all morning. Towels were set out, and dishpans; each man dipped out water to wash with."

OLD-FASHIONED ANISE CAKES

"This recipe was handed down from my grandmother, Hilda Baker Mark (1878–1938), and is just as she gave it," says Betty Carlson. "Can you imagine beating the eggs and sugar by hand? The press used by my mother and grandmother was a rectangular wooden one, with carved designs for marking the cookies. It was made in Germany. Rolling pins with these designs are now available."

MAKES 15 TO 16 DOZEN 1 X 1¹/₂-INCH RECTANGULAR COOKIES

1	dozen eggs	3	level teaspoons hartshorn
3	pounds white sugar		(purchase at drugstore) (see Note)
	Juice of 1 lemon	3	tablespoons water
3	teaspoons anise oil	9½	to 10½ cups all-purpose flour

1. Beat the eggs and sugar together for 1 hour by hand, or 30 minutes with a mixer.
2. Add the lemon juice, anise oil, hartshorn, and water.
3. Beat in flour until the dough is very stiff. Cool, then roll thin (about 1/4 inch) and press with a mold. Cut along imprint lines to separate the cookies. Cover with tea towels and let rest overnight, to rise. Next morning transfer to greased or parchment-lined baking sheets and bake in 350 degree oven about 15 minutes, or until set. The cakes should be very pale on top and light brown on the bottom.

NOTE: Hartshorn is ammonium bicarbonate, used as a leavening agent.

MY MOTHER'S SALAD DRESSING

"We had this often with leaf lettuce from the garden," says Betty Carlson. "At threshing time Mother'd gather a tubful of lettuce for all those men. Our city relatives always asked for 'Mary's dressing.' The same mixture was used on finely chopped coleslaw. After we stopped milking cows and did not have our own thick separated cream, I have made the dressing with commercial heavy or sour cream. It is good, but not the same."

MAKES ABOUT 1 CUP

⅓ cup sugar

¼ to ⅓ cup cider vinegar

⅓ cup heavy or slightly soured cream

¼ cup finely diced scallions (green onions) (optional)

- Dissolve the sugar in the vinegar. Beat in the cream, and add the scallions, if desired.

MARY MARK'S MAYONNAISE

"Mother used her mayonnaise especially for potato salad," says Betty Carlson. "It's really a boiled dressing."

MAKES 1 QUART

3 eggs

¼ cup all-purpose flour

1 teaspoon salt

1 teaspoon prepared mustard

2 cups water

½ cup cider vinegar

1 cup sugar

2 tablespoons butter

1. Beat the eggs in a large bowl. Beat in the flour, salt and mustard. Stir in the water, vinegar and sugar.
2. Put in a large nonaluminum saucepan and cook, stirring, over moderate heat until thick. Stir in the butter and beat until smooth. Chill.

SUMMER QUENCHERS

"My mother made root beer," says Joyce Agnew, "and filled old pop bottles with it. And once a summer, on a very hot afternoon, someone would buy a case of beer. Now, we were a strong Lutheran neighborhood in Darwin, Minnesota, and there was not much drinking. They had that beer only once a summer."

Thrashing, 1889.

MADISON COUNTY FAIR

Winterset, Iowa

FAIR DATE: EARLY AUGUST, 5 DAYS

GAZETTEER: *Attractive rolling wooded countryside, with six scenic covered bridges. Covered Bridge Festival, second weekend of October. Winterset is 13 miles from both I-35 and I-80, at the junction of Highway 169 and Highway 92. The town is 35 miles south of Des Moines.*

MADISON COUNTY SHEEP PRODUCERS
Congratulate all sheep project exhibitors!

Live Show-Tuesday, Aug. 11-8 a.m.
Carcass Show-Thursday, Aug. 13

****
Lambburger Lunch Served
after live shows Tuesday
****

"We're the hub of the world here," says Marlin Brittain, fair secretary and manager. "You can get anywhere from Winterset. We've been famous in Iowa for our fine covered bridges, but now everybody knows us through Robert James Waller's bestseller, *The Bridges of Madison County*. We're also the birthplace of John Wayne. And we're pretty proud of our fair, too!

"We've been a rural 4-H fair, but we hold it on the weekend now, to attract the urban people from Des Moines. We have big tug-of-war contests for the 4-H clubs—basically, it's a way to get kids to release steam so's to get them to bed at night. Our poultry show is growing, and we've added a cat show. We have about a thousand pieces in the 4-H Home Improvement category—welding is a big entry. Sometimes they do sculpture, but basically they just want to show a good weld that will hold.

"We only have five food booths. I could have a hundred, but I keep it down so they all make money. And I keep it to organizations that will put money back into the community. The VFW sells homemade vanilla ice cream, and they wait in line for that. Most people never get homemade ice cream except at the fair. The VFW makes a hundred and fifty gallons, and they always sell out early.

"Horseshoe pitching is a very big thing. We have up to one hundred contestants, and sixteen courts, so thirty-two can go at a time. Some of the best pitchers come, from as far as Nebraska and Missouri. They compete for trophies and glory. The wives come, and bring lawn chairs and knitting. Pitching draws a crowd."

"I'm in charge of the talent show," says Gretchen Brittain. "The talent search is for the state fair, and two winners, from senior and junior acts, go on to the big competition in Des Moines. They give scholarships, so it's taken very seriously.

"We live on a grain and livestock farm. It's a wonderful place to raise a family; clean air, and you are your own boss. The kids figure out what to do on their own. I was a 4-H leader for a good many years, and I usually exhibit at our fair; bakery goods, handiwork, vegetables.

"When I was in high school, back in the Fifties, I took my bedroom to the state fair. I'd refinished the furniture myself, and made the curtains and spread. They set up and painted walls just like mine, and I sat in my 'room' and answered questions about it. Just before I went, the newspaper came to take my picture as I refinished a piece of furniture. I had the mumps at the time, so I had to turn the non-mump side of my face to the camera."

GRETCHEN'S ANGEL FOOD

No cholesterol!

MAKES 1 CAKE

12	egg whites, at room temperature	1½	cups granulated sugar
¼	teaspoon salt	1¼	cups sifted cake flour
1	teaspoon cream of tartar	1	cup confectioners' sugar, for dusting
1½	teaspoons vanilla extract		

- Preheat the oven to 350 degrees.
1. In the large bowl of a standing electric mixer, beat the egg whites, salt and cream of tartar at high speed until frothy; add the vanilla. Gradually add the sugar, beating the mixture until stiff–but not dry–peaks form. Sift in the flour and fold gently until blended.
2. Spoon into an ungreased 10-inch tube pan. Bake 50 minutes or until light brown and springy to the touch and an inserted skewer comes out clean. Turn upside down on a rack and cool for two hours. Run a knife around the rim of the pan and center to loosen. Invert onto a serving platter. Place the confectioners' sugar in a sieve and sift it over the cake.

BUCHANAN COUNTY FAIR

Independence, Iowa

FAIR DATE: THIRD WEEK OF JULY, FOR 7 DAYS

GAZETTEER: *This is flat agricultural country in northeast Iowa, with corn, soybeans, oats, hay and livestock. The fairgrounds are north of Highway 20.*

"Buchanan County is just like Madison County," says Laura Morine, secretary-treasurer of the fair, "except that we don't have bridges. We're the second-largest hog-producing county in the state. I help with our family corporation, that sells twenty-five hundred hogs a year; I work full time in the Farmers Bank, too.

"Our fair has a fruit pie contest, sponsored by the Pork Council; the pies are made from scratch in the 4-H building, and the crust must contain lard, which as you know is a pork product. The pies are auctioned after they're judged, and one could go for as much as three hundred dollars. Two kids might work on a pie, and one set of parents may try to outbid the other; it's a lot of fun, and no hard feelings. Actually, the parents might have pooled their money, and the 'bidding war' is just for show. The money goes to the 4-H.

"The Pork Council used to be called the Porkettes; they were the wives of hog producers who volunteered to promote pork. The older ladies wanted to remain Porkettes, as strictly a women's operation, but the younger ones wanted to bring their husbands in; they felt the men should help with the pork promotion, too. The activities are pretty much the same as before; pork people travel, hit grocery stores

Serious business.

and supermarkets and tell the pork story—how to cook it, and how it is no longer a fatty meat. We also grill pork at weddings, field days and corn festivals.

"North of here is a small Amish community. They never compete in the fair, but the young boys do come down to look—look hard—at our Open Draft Horse Show."

OWEETIS FRYE'S FRESH PEACH PIE

Oweetis was very active in the Porkettes in her younger days, and still keeps up with the pork industry. She is a farmer's widow, continuing to live on the farm; she tends her large vegetable and flower gardens every rain-free day. She bakes from scratch, and the hired men look forward to lunch, when she makes them a full-course meal. "I think Oweetis knows everyone at the fair by name," says Laura Morine, "and just how and to who they're related." The flaky crust of this Pork Council–winning pie is made, of course, with lard.

MAKES 1 PIE

THE FILLING:

4	cups fresh sliced peaches	3	tablespoons tapioca
4	teaspoons lemon juice	¼	teaspoon salt
1¼	cups granulated sugar	2	tablespoons butter

THE CRUST:

2	cups flour	⅔	cup lard
¾	teaspoon salt	4	to 5 tablespoons cold milk or water

- Preheat the oven to 425 degrees.
1. Make the filling: In a bowl, gently toss the peaches, lemon juice, sugar, tapioca and salt together. Set aside for 15 minutes to let the tapioca soften.
2. Make the crust: Combine the flour and salt and cut in the lard with a pastry blender until the largest lumps are the size of small peas. With a fork, mix in enough milk to make a smooth dough.
3. Divide the dough in two and roll out the bottom crust; fit it into an 8- or 9-inch pie plate.
4. Place the filling in the pie and dot the butter over it. Roll out the top crust and place it on the pie, crimping around the edge to seal. Cut decorative vents. Bake for 10 minutes, then lower the heat to 350 degrees and bake for another 30 minutes, or until the crust is nicely browned.

HOT PICKLED GREEN BEANS

Joann Wehner is chairman of the flower show and she has also entered food categories. "The green beans have gone to the fair and won," says Joann. "Good on a relish tray. Very tasty!"

MAKES 7 PINTS

4 pounds fresh young green beans	5 cups cider vinegar
3½ teaspoons hot red pepper	5 cups water
7 heads of dill seed	½ cup pickling salt
7 cloves garlic, peeled	

1. Cook the beans a short time, only until barely tender. Cool under cold running water. Snip both ends of the beans and cut them into uniform lengths.
2. Prepare hot sterilized pint jars (leave lids and sealers in boiling water till needed) and put 1/2 teaspoon of red pepper, a head of dill, and a clove of garlic in each jar. Pack the beans upright, leaving 1 inch of headroom.
3. Boil together the vinegar, water and salt. Pour the boiling liquid over the beans; leave 1/4 inch headroom. Adjust lids and let cool.
4. Adjust the lids again if needed, and let sit for two weeks before using, to let flavors develop.

HOWARD COUNTY FAIR

St. Paul, Nebraska

FAIR DATE: THIRD WEEK IN AUGUST, FOR 4 DAYS, THURSDAY THROUGH SUNDAY

GAZETTEER: *"Square in the middle of Nebraska," Howard County is agricultural. The fairgrounds are in St. Paul, at the junction of Highways 2 and 281.*

"I've been secretary-manager of the fair for over forty years," says Hamie Elstermeier, "because nobody else wants this dirty job. Seriously, though, our fair is one of the biggest donated labor efforts in Nebraska. Roger Welsch, a reporter on the Kuralt CBS Sunday morning show, put our fair on television, to show what a small country fair is like.

"We do most of the judging on Wednesday, and when the fair opens on Thursday, we have a thrill show, stunts with cars and monster trucks, and we've had camel and ostrich races. The minister rode a camel once. Hendrich Shows bring these in. They had zebras to do tricks between the races.

"The Ugly Truck and Pickup contest is real popular. We let the audience fill out slips to pick the winner. We've had some ugly ones! But they have to be runnable.

"Friday is the Demolition Derby, mostly local people. Biggest crowd drawer of the fair. Saturday is children's day, with the Pedal Tractor Pull and the Watermelon Seed Spitting contest. We grow a lot of watermelons in Nebraska.

Reverend Richard Dimond, Methodist minister from St. Paul, Nebraska, in the Ostrich Chariot Race.

It's a good sand area, and watermelons like that. Along the highway you'll find watermelon and muskmelon stands every few miles. Watermelon is served free at the fair.

"Saturday night is the Anything Goes show; a pie eating contest, and stunts like men pushing their 'brides' in wheelbarrows across a bridge built over mud, and of course tipping the gals out. Hilarious things that people like to see their neighbors do, and make fun of.

"The fair was hit by a tornado in April of '84. The fair board said, 'We'll never have a fair this year.' I said, 'Skip it once, and we'll lose the momentum.' So we rebuilt enough to get by that year. We rebuilt solidly later. The tornado took all but two buildings, tore the roof off the grandstand and flattened all the cattle pens. We live between two rivers, and they say that tornadoes don't cross water, but that's nonsense."

GRANDMA NIELSEN'S DANISH CHICKEN SOUP

"This is long because there are really three recipes for my grandmother's soup: the broth, the meatballs and the dumplings," says H. E. (Hamie) Elstermeier.
"I suggest that you flour and brown the boiled chicken and serve it as a side dish, or remove it from the bones and make a chicken salad for a soup and salad luncheon."

SERVES 8

THE BROTH:

1	(3½-pound) chicken, cut into pieces	6	medium potatoes, peeled and cut into ½-inch pieces
2	medium onions, chopped		
6	stalks celery, chopped	2	teaspoons salt
6	carrots, sliced	1	teaspoon celery salt

THE MEATBALLS:

1½	pounds ground beef	1	teaspoon onion salt
½	cup flour	½	teaspoon pepper
1	teaspoon salt	2	eggs

THE DUMPLINGS:

1	cup water	1¼	cups flour
1	teaspoon salt	2	eggs
2	tablespoons butter		

1. Make the broth: Place the chicken in a large heavy pot and cover with water by 2 inches. Bring to the boil, lower the heat and cook, skimming off foam. Cook until very tender. Remove the chicken from the broth.

2. Add to the broth the onions, celery, carrots, potatoes, salt and celery salt. Cook until the vegetables are just tender. Set aside.

3. Make the meatballs: Place the beef in a large bowl. Add the flour, salt, onion salt, pepper and eggs. Mix well. (It's best to use your hands.)

4. To a 6-quart kettle half full of water, add the salt. Prepare the meatballs by wetting your hands and rolling the meat mixture into 3/4-inch balls. Place them on a wet cookie sheet. When the water reaches a rolling boil add a batch of meatballs; let them bob to the top and continue boiling for 1 minute. Remove with a slotted spoon and place in a bowl. Repeat until all the meatballs are cooked. Keep the cooking water to add to the soup broth if necessary.

5. Make the dumplings: In a saucepan bring the water, salt and butter to a full rolling boil; remove from the heat and stir in the flour with a wooden spoon until a stiff ball forms. Add the eggs, one at a time, stirring. (If you want more dumplings, double this recipe.)

6. To a 6-quart saucepan half full of water add the salt and bring to the boil. Prepare the dumplings by wetting your hands and rolling the dough into 3/4-inch balls. Place on a wet cookie sheet, ready to be dropped into the boiling water with a wet spoon. Cook in batches; when the dumplings rise, remove them with a slotted spoon and place in a bowl.

7. When you are ready to serve, heat the broth with the vegetables; add any or all of the meatball-cooking water. Bring to the boil and reduce the heat. Add the meatballs and dumplings and just heat through. Adjust seasoning, if necessary.

MARGE'S MINT JELLY

"I learned the art of jelly making from my aunt in Wyoming," says Marjorie Sperling, "and I married into a family who is big in the county fair. Their thing is horses, but mine is jelly. I've earned many purple ribbons, a few blue and white, but the majority were purples. I've also been superintendent at our fair in charge of the baked and canned goods. I'm always on the lookout for new jelly recipes." Marge says that the jelly is great with lamb, and "good on toast, once in a while."

MAKES 4¹/3 CUPS JELLY

4 cups tightly packed rinsed mint leaves
5 cups water
1 box Sure-Jell or Pen-Jell fruit pectin
   Green food coloring

4 cups sugar
¼ teaspoon butter
   Paraffin for sealing glasses

1. Chop the mint. Place it in a 4-quart heavy pan with the water. Bring to the boil, lower the heat and simmer, uncovered, for 30 minutes. Strain and measure out 3 cups of the mint infusion. It will have a dull, light green color. Discard the leaves.

2. Put the infusion and the pectin in a heavy pan. Bring to the boil and boil 1 minute. Add 3 or 4 drops of green food coloring, or enough to obtain the desired color.

3. Immediately add the sugar and the butter. (The butter cuts down on the amount of foam that rises.) Stir constantly as the liquid comes to a rolling boil, one that can't be stirred down. Boil for 1 minute. Remove from the heat, and quickly skim off the foam.

4. Pour the jelly immediately into hot sterilized jelly glasses to within 1/2 inch of the top. Cover immediately with a 1/8-inch layer of hot, but not smoking, paraffin. Let the glasses stand until the paraffin hardens and the jelly is cool. Cover with metal lids. Store in a cool, dark, dry place.

NEBRASKA STATE FAIR

Lincoln, Nebraska

The ten-day fair, which runs through Labor Day, draws as many as eighteen hundred food entries in a variety of categories. Each year an insert is added to The Nebraska Fair Cookbook. *Below are two of the winners. (For information about the book, write to Nebraska Fair Cookbook, P.O. Box 81223, Lincoln, Nebraska 68501.)*

TUTTI-FRUTTI KISSES

Evelyn Moon, First Place Blooming Prairie Warehouse Natural Desserts, 1991

MAKES 24 TO 30 KISSES

1 cup dates	½ cup ground pecans
1 cup dried apricots	2 cups unsweetened coconut
1 cup raisins	24 to 30 pecan halves

1. Grind the fruit in a meat grinder, or chop it in a food processor, pulsing with the metal blade. Mix in the ground pecans.
2. Form walnut-sized balls and roll them in the coconut. Place a pecan half on top, and flatten slightly.

Pie Ladies.

FISH FILLETS AND FETTUCCINE

Kay Smith, First Place Fish Casserole, 1990

SERVES 4

3 tablespoons flour
1 teaspoon salt
1½ teaspoons lemon pepper
4 to 6 whitefish fillets
2 tablespoons vegetable oil
4 cups cooked fettuccine
½ cup chopped green olives

¼ cup chopped pimiento
½ teaspoon dried thyme, crumbled
1 tablespoon butter
¼ cup chopped onion
2 cups yogurt
½ cup milk
½ cup grated Parmesan cheese

- Preheat the oven to 350 degrees.
1. On a plate or sheet of wax paper, combine 1 tablespoon of the flour, the salt and 1 teaspoon of the lemon pepper. Dip the fillets in the mixture to coat.
2. In a large heavy skillet, cook the fillets in the sizzling oil for 3 to 5 minutes, turning once. Set aside.
3. Butter a 9 × 13 × 2-inch baking dish. In a bowl, mix the fettuccine, olives, pimiento, thyme and the remaining lemon pepper. Transfer to the baking dish and place the fillets on top.
4. In a large skillet, melt the butter and cook the onions over moderate heat, stirring, for 2 minutes. Stir in the remaining flour. Stir in the yogurt and milk. Pour the sauce over the fillets, sprinkle evenly with the cheese, and bake for 40 minutes.

OZARK EMPIRE FAIR

Springfield, Missouri

FAIR DATE: LAST FRIDAY OF JULY, FOR 10 DAYS

GAZETTEER: *The fairgrounds are at the intersection of I-44 and Highway 13. The area is agricultural, with grain farming, dairying and cow-calf operations. To the south is hilly tourist country, including the phenomenal town of Branson, "country music capital of the world."*

"Each year we have a theme for our fair," says Marla Calico, operations manager, "and that sets us apart from other fairs, I think. In 1987 'The Wizard of Oz' was the theme, and we painted the asphalt at the entry to look like the Yellow Brick Road. The complete set of characters—the Tin Man, Dorothy and the rest—wandered around. The year we had SaFAIRi as the theme, we borrowed elephants from the zoo, and people from Kenya set up a booth with African artwork. Fairgoers really get into it, and dress the part. That year lots of them came in khaki shorts. When the theme was the Wild West, they wore boots, denims and Western shirts. In '91, 'Back to the '50s' brought out the poodle skirts and a large Duncan yo-yo contest.

"Hooking people in on a theme is one thing, but when you get them in, you've got to deliver," says Marla. "We have a tradi-

tional fair on a very large scale—over four thousand animals. All breeds of cattle, sheep, even potbellied pigs. To get the farm message out to the urban population, we have a birthing center. We work closely with our dairy farmers, and they bring in cows about to calve. The cow is tended round the clock by agricultural students from Southwest Missouri State University. When she's ready, we announce it on the public address system, and people come to the area and sit on bleachers. They are quiet and reverent; it's so inspiring. Mothers look at it as an opportunity to teach kids about life. I grew up on a dairy farm, and I hadn't realized just how special this is to see. We use dairy cattle because they are used to being handled by humans on a daily basis, and don't get upset. Farmers schedule when their cows calve. Cows have a cycle; dry before they give birth, and lots of milk afterward. So if you are in business, you want to keep that milk coming on a steady basis."

"I used to be in the A&W Root Beer drive-in business," says general manager Dan Fortner, who had two of the largest stands in the country. "My sons run the concession here, and we're known for our corn dogs and root beer in quart jugs.

"We have twelve thousand entries in our household arts. When I took over in 1982, entries had gone down to seven thousand. I started urging everyone to enter, get the numbers up. I talked my wife into entering, and darned if she didn't win Grand Champion with her custard pie! I was really in a state that people'd think the contest was rigged, But then I said, "The best pie is from Missouri, and the judges proved it's hers."

LOVENA FORTNER'S CUSTARD PIE

"The pastry recipe makes enough for three single crusts," says Lovena. "You'll only need one for the pie; freeze the rest for another day."

MAKES 1 PIE

THE PASTRY:

3 cups flour	1 large egg
1 teaspoon salt	1/3 cup cold water, plus a few more
1 tablespoon sugar	drops, if needed
1 1/4 cups solid vegetable shortening	1 tablespoon cider or white vinegar

THE FILLING:

4 large eggs	1/4 teaspoon freshly grated nutmeg
2/3 cup granulated sugar	2 2/3 cups milk
1/2 teaspoon salt	1 teaspoon vanilla extract

- Preheat the oven to 450 degrees.
1. Make the pastry: Use a pastry blender to mix the flour, salt and sugar. Cut the shortening into the mixture until it resembles coarse meal.
2. In a separate bowl, whisk together the egg, cold water and vinegar. Add to the flour mixture and blend until a dough is formed. Add a little more water, if necessary. Cut off one-third of the pastry and roll it out to fit a 9-inch pie plate.
3. Make the custard filling: Beat the eggs slightly with a rotary or electric beater. Beat in the sugar, salt, nutmeg, milk and vanilla. Pour through a strainer into the pie plate. Flute the edge.

4. Bake the pie in the center of the oven for 20 minutes. Reduce the heat to 350 degrees, and bake for 20 to 30 minutes longer, until a knife inserted halfway between the center and the edge comes out clean. (To keep the edge of the crust from burning, cover it with strips of aluminum foil as the custard bakes.)

HAZEL BARNETT'S FIG JAM

Hazel has worked in the Ozark Empire Fair office since 1980. She makes this jam for her coworkers every fall. "Great with biscuits," she says.

MAKES 5 PINTS

2 quarts fresh figs

¾ cup water

6 cups granulated sugar

¼ cup freshly squeezed lemon juice

1. Put the figs in a large bowl and cover with boiling water. Let stand for 10 minutes. Drain and, when cool enough to handle, stem and chop them.
2. Put the figs in a large nonaluminum pan and add the water and sugar. Slowly bring to the boil, stirring occasionally until the sugar is dissolved. Raise the heat and cook until thick, about 1 hour, stirring frequently to prevent sticking. Add the lemon juice and cook 1 minute longer.
3. Pour, boiling hot, into hot sterilized jars. Process in a boiling water bath for 15 minutes.

FORD COUNTY FAIR

Dodge City, Kansas

FAIR DATE: LATE JULY, EARLY AUGUST, FOR 4 DAYS

GAZETTEER: Made famous by Gunsmoke *and other television dramas, Dodge City is a thriving town of feedlots and meat packing plants. One of its attractions is the Boot Hill Museum, a preserved and reconstructed collection of structures housing artifacts of the 1870s. The reconstructed mercantile buildings and saloons on nearby Front Street further re-create the era when Dodge City was the terminus of the Chisholm Trail, Bat and Ed Masterson kept the law, and five million longhorn steers were shipped east on the railroad. The fairgrounds are on Park Street, south of Wyatt Earp, the main street.*

"Our fair is part of Dodge City Days, a ten-day community celebration, with a rodeo, a chuck wagon breakfast, pit barbecue and Western parade," says Diane D. McNeill, county 4-H Extension Agent.

"We have lots of activities for the kids, including the Turtle Races. These are land turtles that people find around here. We usually have over one hundred turtles, and we break them into groups for the races. We also have a free ice cream social and that really draws the crowds.

"As you know, Kansas is the Sunflower State, so we have a Sunflower Project. In the spring we give out over twenty-six hundred packets of sunflower seeds to schoolchildren, kindergarten through fourth grade, to plant and watch grow. The sunflowers grow eight to twelve feet tall, with heads that can measure up to twenty inches across. Like Jack in the Beanstalk! On the first day of the fair, the children cut the flowers at the base and bring them in in the morning. That evening, prizes go to the Tallest Plant and the Biggest Head. Every child who enters, though, gets a ribbon and a coupon for a free drink."

"At the county fair I ran a Horticultural Ident Program," says Garrett McClure, a high-schooler who has been in 4-H since he was nine and is president of the Richland Boosters 4-H Club." I displayed twenty-four vegetables and seeds, and twenty-seven people took part in my contest. My main 4-H projects are Leadership, Gardening, and Food Preservation. Gardening and food preservation projects go well together; you grow the produce, then you preserve it for the winter.

"I live with my parents and two brothers on a farm. My

Matt, Briana and Anne Durler, with a Kansas sunflower.

whole family helps with the chores of canning. This is a big help when it comes to snapping beans or blanching corn. Recently I preserved seventy-five quarts of green beans and fifty-six dozen ears of sweet corn, and a whole lot else. This was the first time we took the corn off the cob. We tried a knife and a potato slicer, but we discovered the best thing to use is an electric knife. It didn't make so much cream as a regular knife, and it was easier to get close to the cob without cutting into it. I sell a lot of my vegetables at the Farmers' Market, and share them with nearby elderly people and with my grandmother. I use only organic methods to raise produce, and I choose seeds that produce ears of corn with husks so tight I don't have much trouble with worms. I don't have much trouble with rabbits, either. Our dog T-Bone gets all of them.

"Photography and art are interests of mine, and I play the cello in the community orchestra. I really enjoy cooking and baking, too."

Garrett McClure.

GARRETT'S CARROTS (AND ASPARAGUS)

"I enjoy looking through cookbooks," Garrett says. "I don't know if cooking will be a career, but it is a pleasure."

SERVES 6 TO 8

2 pounds carrots, peeled, quartered and cut into 1½-inch sticks	½ teaspoon freshly grated lemon peel
1 pound asparagus, ends trimmed, cut into 1½-inch pieces	2 tablespoons freshly squeezed lemon juice
2 tablespoons butter	½ teaspoon salt
	½ teaspoon freshly ground pepper

1. Bring 1 inch of water to the boil in a large deep skillet. Add the carrots, cover and simmer over moderate heat until crisp-tender, 4 to 6 minutes.
2. Remove the carrots with a slotted spoon to a colander placed in the sink. Immediately run cold water over the carrots to cool them. (This keeps them from cooking further, and preserves the color.)
3. Add the asparagus to the skillet, cover, and simmer 6 minutes, or until crisp-tender. Tip into the colander.
4. Melt the butter in the same skillet. Add the lemon peel, juice, salt and pepper. Add the vegetables. Stir 1 to 2 minutes or until hot. Serve at once.

GARRETT MCCLURE'S NEW POTATOES AND PEAS IN CHEDDAR SAUCE

SERVES 8

2 pounds new potatoes, cut into 1-inch chunks (about 7 cups)

3 cups chicken broth

2 tablespoons butter

2 tablespoons flour

2 teaspoons dry mustard

⅛ teaspoon ground hot red pepper

1 teaspoon salt

1 cup milk

2 cups defrosted frozen or cooked fresh peas

4 ounces shredded cheddar cheese (1 cup)

1. Bring the potatoes and broth to the boil in a 3-quart saucepan. Cover, reduce the heat to low, and simmer 15 to 20 minutes, until the potatoes are tender, but still firm. Drain, and reserve 1 cup of broth for the sauce.

2. Make the sauce: Melt the butter in the same saucepan, add the flour, dry mustard, red pepper and salt, and whisk over low heat until smooth and frothy. Let the mixture bubble for about 3 minutes, stirring often to prevent browning. Gradually whisk in the cup of broth and the milk. Increase the heat to moderate and whisk the mixture for 2 to 3 minutes until the sauce thickens.

3. Add the peas to the sauce and bring to the simmer, stirring gently. Add the potatoes and stir gently to coat. Reduce the heat to moderately low. Cover and heat 5 minutes, stirring once or twice. Remove from the heat, and stir in the cheese. Serve at once.

Norval Lembright with his 4-H Club project, 1940.

ROOKS COUNTY FREE FAIR

Stockton, Kansas

GAZETTEER: *Stockton is forty miles from Fort Hayes, at the intersection of Highways 24 and 183.*

PREMIUM LIST OF ROOKS COUNTY FAIR.

BASE BALL GAMES.

Arrangements are being made for a good game of base ball during the fair between the best clubs in this part of the country. The following purses are offered:

Wednesday, Sept. 7, 1910

The Winner	$20.00
The Loser	10.00

Thursday, Sept. 8

The Winner	$40.00
The Loser	20.00

Friday, Sept. 9

The Winner	$20.00
The Loser	10.00

W. T. PFLEIDERER

AUCTIONEER

Phone 250, or call at Auction Room in Phelps' Building.

STOCKTON, KANS.

From the 1910 premium book.

"I think it's of interest that we had a fair before we had a town," says fair secretary James Ochampaugh. "Stockton was government land until 1880, but the settlers organized an agricultural society and held a fair in 1879. The fair brought everyone together, and it still does; makes them think as one. It's a driving force to organize the community, and to exhibit our wares. It's our heritage. I was born in 1941, joined the 4-H and grew up thinking 'Fair.' Still do.

"This is cattle country. Rooks County has plenty of cattle, but a human population of fewer than seven thousand. We have what we call old timers' livestock judging at the fair. I judged in high school, and I've gone back to see if I still have skills. Lots who used to judge are still very qualified. This judging doesn't count on the animals' records. It's strictly the judges being judged.

"We have a Youth Pet Parade, for the little brothers and sisters of the 4-H'ers. They show up with chickens, puppies, cats, baby calves, and even pot-bellied pigs. The emcee talks to each child, who has a little time on display. The parade is in the show area, and draws about forty-five hundred people. Each child gets a yellow ribbon.

"A new category at the fair is wine. Before me, George Ostmeyer ran the fair for thirty years, and wine making was his hobby. After he retired, he started giving wine classes. We had twenty-five entered recently.

"Up through the 1970s, every town around here had an American Legion baseball team. They had playoffs at the fair, right in the infield. While the teams were playing, motorcycle races were going on around the track. The track was originally for horse racing, but in 1952 we added professional motorcycle racing. National dirt track racers acquire points at tracks like Rooks. I recall one racer, a hundred-pound star, called the Flying Flea. Harness racing ended in 1989—not fast enough for the young."

"I taught home ec in school, so I didn't think it was appropriate for me to exhibit then," says Eda Jean Hildreth, "but I've done a lot of judging of preserved food and jellies. My daughter Raenetta was very active in 4-H, exhibiting pigs, cakes, and clothing.

"My family came to Kansas in 1874 in a wagon train. My great-grandfather had to go nearly a hundred miles from his place to the railroad to get supplies. Once while he was gone, his wife set backfires in the field to scare the Indians away. Some of the family moved back and forth to the Cherokee Strip in Oklahoma. They led a cow that went dry on the trip, so no milk for the children. They looked for shade trees to camp under. The mother and five children slept in the wagon, and the father slept underneath. My mother, who lived to be ninety-five, was on that trip. Later on, my mother flew over Oklahoma in a jet plane. She had gone there as a child in a covered wagon, and she lived to fly over the same route."

EDA JEAN HILDRETH'S
WHEAT BERRY CHILI SOUP

"Here in Rooks County we don't have to ask 'Where's the beef?'" says Eda Jean.
"We know where it is! And as Kansas, the wheat state, is the breadbasket of the world,
I'm including a recipe with wheat berries, which we use a lot."

SERVES 8 TO 10

3 cups cooked wheat berries (see Note)	1 cup chopped fresh or canned tomato
3½ pounds ground chuck or other beef	1 (15-ounce) can tomato sauce
1 cup chopped onion	2 cups water
3 tablespoons chili powder	1½ teaspoons salt
3 tablespoons chopped green pepper	

1. To prepare 3 cups of cooked wheat berries, rinse 1 cup of dry berries. Place in a large saucepan and add 2 cups of boiling water, salted to taste. Let stand overnight. Next day bring to the boil, lower the heat and simmer for about 30 minutes or until tender. (The berries should retain a bit of crunch; boil them too long and they turn to mush.) Berries can also be prepared ahead of time and frozen.
2. In a large heavy nonaluminum saucepan, cook the meat over moderate heat, pressing down with a wooden spoon, until the meat has browned slightly. Add the onion and the chili powder. Cook, stirring, for 3 minutes. Add the green pepper, tomatoes and tomato sauce, water and salt. Partially cover and simmer for 45 minutes. Ten minutes before the soup is ready, add the wheat berries and, if needed, add extra water.

NOTE: Wheat berries—whole unprocessed wheat kernels—can be found in health food stores. They can also be ordered from The Granary, 1303 Third Avenue, Downs, KS 67437.

SEVEN

THE WEST

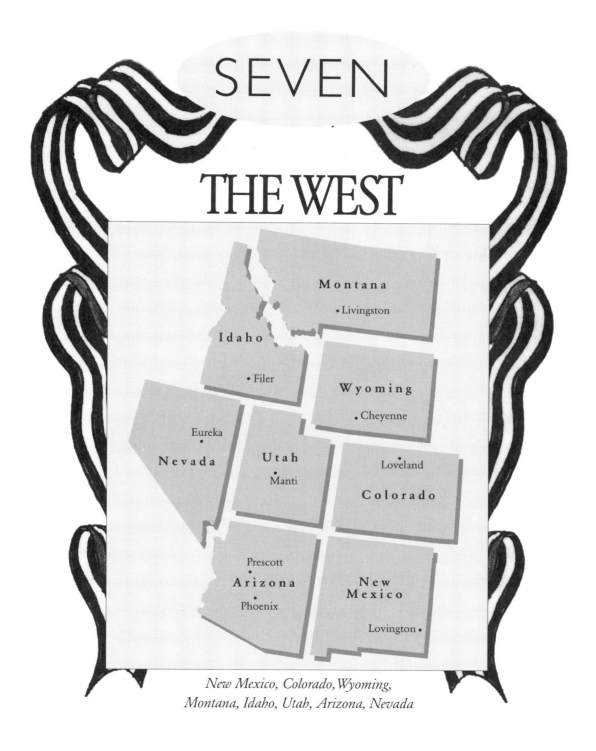

Montana
• Livingston

Idaho
• Filer

Wyoming
• Cheyenne

Eureka
•

Nevada   Utah
• Manti

Loveland
•

Colorado

Prescott
•
Arizona   New
• Phoenix   Mexico

Lovington •

New Mexico, Colorado, Wyoming,
Montana, Idaho, Utah, Arizona, Nevada

LEA COUNTY FAIR & RODEO

Lovington, New Mexico

FAIR DATE: SECOND WEEK IN AUGUST, FOR 8 DAYS

GAZETTEER: *The Lea County fairgrounds are on State Highway 18.*

Lea County, in the southeast corner of New Mexico, nestles into the Texas Panhandle; the country was, in fact, once part of Texas. This is cattle-raising country. Ranches are counted in sections, each section being six hundred and forty acres. Ranching is hard, slogging, day-in, day-out work, three hundred and sixty-five days a years, so ranchers must stick close to home—except for the young men who choose to leave and go on the rodeo circuit. Lea County boasts more world-champion rodeo performers than any other county in the United States. "Home of the Rodeo Champions" is the motto of the annual fair and rodeo, and come home the ropers and riders do, to compete for rich prizes in the largest pro rodeo in the state. Lea County's population is seventy thousand, and for the fair, over a hundred thousand visitors come.

The rodeo begins nightly at the Jake McClure Arena, but agricultural events dominate the daytime. The first Saturday begins at eight a.m., with simultaneous events: the 4-H and Future Farmers of America horse show; 4-H'ers and FFA'ers shearing their lambs; rabbit and poultry events, and a Mexican rodeo.

"We'd bring our livestock to the fair and spend a week in the camper," says Stacy Medlin Reid, a young rancher who competed as a 4-H'er. "We'd feed them, wash them, show them, and on Friday we sort of took a day off. We'd have a huge water fight—kids and adults, too. The fairground has about forty water hoses people use to wash animals, and we'd turn the hoses on each other. There's a special concentrated shampoo used for calves and horses, so we threw that around and it takes days to get out of your hair. The same people come year after year. You make a second family at the fair."

Medlins have been in Lea County before it had a name, since Christopher Columbus Medlin, a sometime army scout and buffalo hunter, found two shallow streams and dug out wells that allowed him to raise cattle. That was in the early 1880s. His son, Hawk, enlarged the property around 1905, buying the eight-section Adobe Place for "$25 and a black horse." He could run as many cattle as he could water on those twenty-five hundred acres—about three hundred head.

Buddy Medlin, Stacy's grandfather, was born here in 1907, in a half-adobe, half-dugout house. He married, bought another ranch, and then a few more. The Medlin Cattle Company is now one of the largest ranches around: sixty-four thousand acres, worked by the extended Medlin family, including Stacy and her husband, Jace Reid. Buddy died in 1985. In 1991, he was posthumously inducted into the Lea County Cowboy Hall of Fame, honored as "an active member of the New Mexico cattle growers, outstanding 4-H leader, an organizer of both a roping club and the Lea County Open Range Cowboy's Association."

"To us, 'cowboy' doesn't mean rodeo," says Stacy. "A cowboy works and tends cattle for a living. Buddy was as fine a roper as ever lived, but his life was ranching. Now my cousin Jeff Medlin is a champion rodeo roper, and I bet he's on the road three hundred days a year—California one night, next night he may be in Arizona, then on to Idaho. Probably fewer than two percent of rodeo cowboys really ranch."

The ranchers here "neighbor," or help out, one another. "In spring," says Stacy, "instead of paying cowboys for branding calves over three days, we neighbor, and everyone comes. Cowboys on branding days come here about four a.m., eat a big breakfast, and as soon as the sun comes up start gathering calves. It might take five hours or more to bring them in. You need about twenty-five people to separate the cows and calves in the pen, some on horseback roping and dragging the calves up to the open fires and the branding iron, and others vaccinating. When my grandfather Buddy was seventy-six, he was roping in the pen, got tripped up and broke two ribs. Next year he was back in the saddle, right out there."

The day's work is through by noon, so ranchers can go home and tend to their own cattle. "No matter how long the job, nobody works past noon," says Stacy. "In a typical season, our own branding plus going to neighbors, folks might be gone every morning for a month. It's a community thing."

While the men whoop, rope and brand, the women cook the midday dinner. Beans, bread, salads and roasts are popular, and so is Mexican food. "We eat a lot of Mexican food down here," Stacy says.

The Medlins. (Stacy is second from right.)

DONNA MEDLIN'S POSOLE (HOMINY STEW)

"Just as Southerners eat black-eyed peas on New Year's Day for luck, we eat posole to bring luck in the New Year," says Stacy Medlin Reid. "This is one of my mother's recipes."
Posole is a savory, satisfying stew based on the dried, skinned and soaked corn kernels called hominy. New Mexicans and other Southwesterners can buy two-pound bags of prepared posole (see Note below), but elsewhere it is sold in cans.

SERVES 6 TO 8

1 tablespoon lard or vegetable oil	1 tablespoon dark brown sugar
2 pounds boneless pork shoulder, cut into ½-inch cubes	1 tablespoon salt
	1 teaspoon crumbled dried oregano
1 cup chopped onion	6 medium-hot red chiles, chopped
3 garlic cloves, peeled and chopped	1 (1-pound) can white hominy, drained
2 tablespoons chile powder	1 (1-pound) can yellow hominy, drained

1. In a large heavy casserole, heat the lard over moderately high heat and brown the pork cubes in batches. Remove from the casserole and set aside. Add the onion to the fat remaining in the pan, along with the garlic, chile powder, sugar, salt and oregano. Cook briefly, stirring. Stir in the chopped chilies. Return the pork cubes and their juices to the pot, and cover the contents with water.
2. Bring the mixture to the boil, lower the heat and simmer for about three hours. Add the white and yellow hominy and heat through. Serve with lime wedges and hot buttered tortillas.

NOTE: Donna Medlin uses prepared posole from local markets, adds it at the beginning and cooks the stew for four hours.

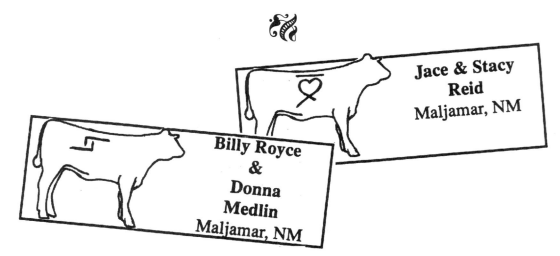

Jace & Stacy Reid
Maljamar, NM

Billy Royce & Donna Medlin
Maljamar, NM

DONNA MEDLIN'S GREEN CHILE STEW

*Donna is Stacy's mother, and an excellent cook. Triple this for a "neighboring" meal!
The medium-hot to hot New Mexican green chile called for also goes under the names Ana-
heim, Big Jim and chile verde. It is bright green, four to eight inches long, and resembles an
elongated bell pepper. Substitute a combination of mild green chiles and jalapeños.*

SERVES 6 TO 8

1 pound coarse-ground pork	Pepper, to taste
1 pound coarse-ground beef	2 teaspoons dried oregano
1 cup canned tomatoes, with juices	2 teaspoons ground coriander
2 teaspoons garlic powder (optional)	4 pounds potatoes, peeled and cut into
1 head garlic, chopped	1-inch dice
1 teaspoon salt, or to taste	8 to 10 green chiles, chopped

1. In a large, heavy casserole, brown the pork and beef over moderate heat, stirring often. Pour off most of the fat.
2. Add the rest of the ingredients, stir well and cover with water. Bring to the boil, lower the heat and simmer, partially covered, for 45 minutes. Correct the seasonings and serve with bread or tortillas.

BUDDY MEDLIN'S BEEF JERKY

*"Cowboys always carried beef jerky with them," says Stacy. "It lasted six months or so.
My grandfather Buddy Medlin made it all the time—I remember strips of jerky hanging on
the clothesline."*

MAKES ABOUT 1 POUND

3 pounds round steak or other lean beef	1/3 cup ground black pepper
1 cup salt	

- Cut the beef into 1/4-inch strips across the grain. Combine the salt and pepper and roll the strips in it. Hang on a line in an airy shaded place, such as a screened porch. (Nylon twine makes an excellent line.) Jerky is ready to eat when brittle and dry.

TRAIL BEANS

Donna Medlin's father, Herbert Price, was born in 1905. He well remembers the chuck wagon, and recalls that it was used until the early 1930s.

Cowboys come home at night now, but in the old days they stayed out on the range for days on end. Along with the cowboys went the chuck wagon, provisions and the cook, called Cookie. Biscuits, beans and beef were the staples, and the cook used a Dutch oven, a heavy cast-iron pot with three legs to raise the bottom above a bed of coals. The heavy lid had a rim designed to hold more coals, so the food cooked from the top as well as the bottom. Getting the food right was a matter of skill and guesswork; no peeking under the red-hot lid.

Mr. Price fondly recalls beans like these. "Easier these days, on the stove," he says. For sentimental reasons, he still has the old Dutch oven.

SERVES 8

2 cups (1 pound) dried pinto beans
6 cups water
   3-ounce piece of lean salt pork,
     rind removed

4 dried hot red chiles about 4 inches
   long, stemmed and crushed
1 medium onion, sliced thin
2 teaspoons salt

1. Rinse the beans under cold running water; discard any blemished beans. Put the beans in a large pot and cover with water by 2 inches. Bring to the boil, boil 2 minutes, remove from heat and let soak 1 hour.
2. Drain and rinse the beans and return them to the pot. Add the 6 cups of water, the salt pork, the chiles and the onion. Bring to the boil, reduce the heat and simmer the beans for five hours, covered. Check from time to time; add boiling water if necessary. (Cold water toughens the beans.)
3. Stir in the salt and simmer for 30 minutes, or until a bean can be easily mashed against the side of the pot. The beans should have absorbed most of the liquid. Serve with a fresh tomato salsa, if desired.

FIRST, CATCH YOUR RATTLESNAKE

Ingenious cooks will use whatever is to hand. Stacy owns an old volume called *The Ranch Cookbook* by the Ladies of the Ranch, in which this recipe appeared.

DELICACY OF DIAMONDBACK

1	western diamondback rattlesnake		Salt and pepper to taste
1	cup flour	½	teaspoon garlic powder
½	cup crackermeal		

- Find and kill a western diamondback rattlesnake. Skin and remove entrails. Cut into edible portions. Combine other ingredients for coating mixture—more may by required, depending on the length of the snake. Coat the pieces, and fry in deep fat until golden brown.

LARIMER COUNTY FAIR & RODEO

Loveland, Colorado

FAIR DATE: EARLY AUGUST FOR 8 DAYS, WEDNESDAY TO WEDNESDAY

GAZETTEER: *Larimer is "front range" country, in the foothills of the Rockies. The area grows winter wheat and sugar beets. Loveland is 45 miles north of Denver, 15 miles north of Boulder. The fairgrounds are off South Lincoln Avenue, U.S. 287.*

"We've got the largest free mule show in the West," says fair director Bob Holt. "They're judged just like horses; we have obstacle courses, pulls, and they are judged on conformation. It's an all-day event.

"We have a huge free pancake breakfast on Sunday morning, and after that there is a church service on our stage, run by the Fellowship of Christian Cowboys.

"A while back we had a Zany Zucchini contest— squash carved and dressed up to look like a bride and groom, or birds, things like that. One girl submitted a zucchini sheriff beating on a small victim squash; there was some misunderstanding and she took it as an insult, or a disqualification; she called the newspapers, and this fuss even got on *Good Morning, America*! It was all slightly embarrassing.

"We have a great rodeo, and performances by the Senior Rodeo—that's kids in high school, and the Junior Rodeo—for younger kids. We have a small arena where you sit right on top of the performers, and you can see the rodeo like it should be seen."

"I hate chili, but I won the Chili Cookoff two years in a row," says Connie Strack. "Of course, one victory was by default. I was the only one who showed up!

"When my kids Stephen and Tiffany were smaller, they did rabbits for 4-H. We once had about seventy of them, and we sold one for seventy-five dollars, for breeding stock. We've only kept one, a Rex, who is marked like a calico cat. She's our pride and joy. She will roll over and stay, and even use the litter box.

"Some people get mad because I bring about sixty-five entries to the fair. My friend Mona Hansis and I cook so much that we said, 'Let's do booklets and sell our recipes!' We advertise in the classifieds in the back of magazines like *Women's Workbasket*. I'd like to do diabetic recipes; diabetes runs in my family, and I'd like to avoid it by eating right. I keep country values—we all rotate chores, and my husband and son can cook. Tiffany's been cooking since she was six, and she's got the awards to prove it."

Lori and Tiffany, with ribbons.

Bull rider, 1993 Fair.

LITTLE THOMPSON 4-H CLUB SLOPPY JOES

"When they were eight and nine, my daughter Tiffany and her friend Lori Riggins won the Creative Cooks Junior Division with their sloppy joes, at both the county and the state level," says Connie Strack. "They had to set up their table, make the dish, serve it to the judge, and answer questions about the nutrition of the menu they had created. Tiffany and Lori dressed alike. The mothers were so proud!"

MAKES 4 SERVINGS

1½ pounds hamburger
1 medium onion, chopped
1 tablespoon Morton's Nature
  Seasoning Blend
½ cup K. C. Masterpiece Original
  Bar-B-Que Sauce

1 (10¾-ounce) can tomato soup
1 tablespoon yellow mustard
2 tablespoons dill pickle juice
4 buns

1. Place the hamburger, onion and Nature Seasoning in a large heavy skillet. Cook over moderately high heat until the meat is no longer pink and has begun to brown. Drain off the grease.
2. Add the K.C. sauce, tomato soup, mustard and pickle juice and mix well. Simmer over moderate heat for 10 to 15 minutes. Serve over buns.

CONNIE STRACK'S ZUCCHINI BREAD

"A prize winner every time," says Connie.

MAKES 2 LOAVES

3 eggs, beaten
1 cup vegetable oil
1½ cups sugar
3 cups grated zucchini
2 teaspoons vanilla extract
1 teaspoon ground cinnamon

1 teaspoon salt
3 cups all-purpose flour
1 teaspoon baking soda
1 teaspoon baking powder
½ cup chopped dates or nuts, if desired

- Preheat the oven to 350 degrees.
1. Mix the eggs, oil, sugar, zucchini and vanilla in a large bowl. Stir in the cinnamon, salt and flour mixed with baking soda and baking powder. Stir in the chopped dates or nuts, if you are using them.
2. Divide the batter between 2 greased and floured loaf pans. Bake in the center of the oven for 1 hour, or until a toothpick inserted in the center of a loaf comes out clean.

SPONSOR

BAXTER BLACK, COWBOY POET

He speaks for the farmer, the rancher, and all who make their livelihood in agriculture. His column, "On the Edge of Common Sense," runs in more than one hundred publications. Baxter appeared frequently on the Johnny Carson show, and is a regular on National Public Radio. Coyote Cowboy Company, his own concern, has published ten books, numerous tapes and a video. He doesn't believe in agents or contracts. His work is free of meanness and profanity, though his humor can be earthy. The man gets around. He makes over one hundred banquet and speaking appearances a year, and recites his poetry at fairs. They sure love him in Sanpete and Tulelake.

Baxter Black was a veterinarian for thirteen years, working for ranches and feedlots in the mountain West. "He was constantly at ranches way out in the boonies," says his wife, Cindy, "and at night there was often no electricity, no TV. They had to entertain themselves. He found that with his poetry he could tease crusty old ranchers without makin' 'em mad. It's amazing how many women come up to him and say, 'You know my husband by heart.'" Baxter decided to quit the vet business, and became what one reporter called "Will Rogers' weird grandson."

Baxter says of those days on the range, "A person can stand only so much guitar playing. Cowboys with terrible singing voices began reciting song lyrics, then they made up their own poetry.

"Rural America is pretty honest," he says. "Rural values stem from the care of animals and nature, as well as from tradition and religion. There's something intuitive about the care. When I feed my horses, I don't think of it as putting gas in my car."

As he regularly picks on cowboys, sheepherders, bad horses and belligerent cows, it's not surprising the animal rights fanatics and vegetarians get skewered with the same pen, as witnessed in his poem "The Vegetarian's Nightmare." Some vegetarians, too, are said to appreciate it.

Baxter Black and equine friend.

The following are selected verses from "The Vegetarian's Nightmare" (subtitled "A Dissertation on Plants' Rights") by Baxter Black:

I had planted a garden last April
And lovingly sang it a ballad.
But later in June beneath a full moon
Forgive me, I wanted a salad!

So I slipped out and fondled a carrot
Caressing its feathery top.
With the force of a brute I tore up the root!
It whimpered and came with a pop!

Then laying my hand on a radish
I jerked it and left a small crater.
Then with the blade of my True Value spade
I exhumed a slumbering tater!

I butchered the onions and parsley.
My hoe was all covered with gore.
I chopped and I whacked without looking back
Then I stealthily slipped in the door.

My bounty lay naked and dying
So I drowned them to snuff out their life.
I sliced and I peeled as they thrashed and they reeled
On the cutting board under my knife.

Then I took the small broken pieces
I had tortured and killed with my hands.
And tossed them together, heedless of whether
They suffered or made their demands.

I ate them. Forgive me, I'm sorry
But hear me, though I'm a beginner,
Those plants feel pain, though its hard to explain
To someone who eats them for dinner!

I intend to begin a crusade
For PLANTS' RIGHTS, including chick-peas.
The ACLU will be helping me too.
In the meantime, please pass the bleu cheese.

CINDY AND BAXTER BLACK'S SAUCE FOR RIBS AND "GRITS"

"No," says Cindy, "not the cornmeal grits you're thinking of. Our grits are sandy Colorado soil, loosened at dusk by a fierce wind that starts west of the Continental Divide! The music group Riders in the Sky did a video with Baxter, and they were staying with us. They wanted a barbecue. We have a nice little camp away from the house, with a small barbecue pit level to the ground. The ribs were covered with sauce, and then the wind kicked up. We're used to that, but it was a major *wind that blew half the food off the table. The grit blew all over the ribs, and the wind fanned the coals to white hot so they charred the food; some said to give it up, but we hosed the ribs down, made new sauce, and managed to eat them. They were actually quite good! But we were in hysterics because they were really gritty.*
"Enjoy ribs without grits. *The recipe was given to me by Cristy Meyer from Gothenberg, Nebraska, and appeared in a book by two gals who are wives of rodeo cowboys, called* Come an' Git It 2. *"*

MAKES ABOUT 2 CUPS

1 cup ketchup	1 teaspoon salt
¼ cup vinegar	¼ cup fresh lemon juice
3 tablespoons Worcestershire sauce	1 small can tomato paste
1 teaspoon paprika	1 tablespoon brown mustard
¾ cup packed brown sugar	

- Mix all ingredients together. Use the sauce to marinate 3 to 4 pounds of pork spareribs for several hours, or overnight if possible. Grill, or cook in a 350 degree oven until tender.

SWEETWATER COUNTY FAIR

Rock Springs, Wyoming

FAIR DATE: FIRST WEEK OF AUGUST, FOR 7 DAYS

GAZETTEER: *Rock Springs, in southwest Wyoming, is near the Red Desert, an area of moving sands. Pictographs and petroglyphs found on canyon outcroppings point to human habitation over six thousand years ago. The Red Desert is home to one of the largest herds of wild horses in the nation; captured mustangs may be adopted at the Wild Horse Holding Facility in Rock Springs. Take the Elk Street exit from West I-80, drive one half mile to the fork, then two miles to the fairgrounds.*

"Rock Springs was a rowdy town in its early days," says Karen Bonomo, fair administrative assistant. "Butch Cassidy, who later robbed banks with the Sundance Kid, lived here. He was a butcher; that's how he got his nickname.

"Later Rock Springs became a coal mining town, and labor demands drew people from all over. The Chinese came, and people from Poland, Germany, Italy—all of Europe. We only have about twenty thousand people in town, but they represent fifty-seven nationalities! Makes for an interesting community.

"Each year we pick an animal theme for the fair. One of my favorites was 'The Year of the Pig.' We trained our own racing pigs, made little silks and caps for them, and let 'em out of a chute to race for cookies. We had a hog calling contest, and pig wrestling in mud. The businessmen came out for that, and lost every time. The 'Wild N Woolly West' year featured a lamb stew cookoff and sheepherders' stories. Wyoming is a big wool state, and there are a lot of sheep around here. At the 'Hen and Rooster Show' fair we had a Chicken Soup Cluck-off for the best recipe, and the Flock-Around-the-Clock Marching Band.

"The Ice Cream Freeze-off goes over very well on a hot Saturday afternoon, believe me. This is hand-cranked ice cream, churned for several hours. We use local celebrities; we furnish the cream, they do the flavorings—chocolate, even spirits. The audience gets little samples. The contestants really ham it up, and that's why I love media people.

"Wyoming Game and Fish stocks the ponds at the fairgrounds with hundreds of rainbow trout just before Kid's Day. We have a Fishing Derby for boys and girls under twelve. Trophies and ribbons are given, and they get to keep the fish.

"We have a lot of choice at our concessions. Slovenian smoked sausage on a

bun, called kronski, is very popular here, and so are tacos and barbecued beef. I think the secret to a good fair is the good food."

"I'm a fair supporter, and I was a 4-H leader for fourteen years," says Frances Koler. "I've had a catering service for quite a few years, and I do a lot of weddings—Slovenian, Swedish, Croatian, even Basque—you name it. Most people came originally for the coal mining, but the Basques are famous sheepherders; that's why they came here."

Palomino halter class.

WYOMING BASQUE PAELLA

"The saffron gives a deep, attractive yellow color," says Frances. "If you cannot get chorizo, substitute the smoked garlic sausage called kielbasa."

SERVES 8

1 (3-pound) chicken, cut into serving pieces, or the equivalent in parts
½ cup olive oil
2 cups raw rice
½ teaspoon saffron dissolved in
¼ cup chicken broth
1 cup finely chopped onion
3 garlic cloves, finely chopped

4½ cups chicken broth
1 cup canned pimiento, cut into strips
2 tomatoes, peeled, seeded and chopped
1 chorizo sausage cut into ½-inch pieces
1 cup frozen peas, defrosted
1½ pounds fresh shrimp, shelled and deveined

• Preheat the oven to 350 degrees.
1. In a large heavy pot, brown the chicken well in the oil over moderate heat, turning with tongs. Lower the heat and let the chicken cook for 10 minutes.
2. Add the rice, saffron, onion, garlic and chicken broth. Bring the mixture to then simmer and let cook for 10 minutes. Add the peas and shrimp. Cover and bake for 25 minutes, or until the rice is thoroughly cooked.

SLOVENIAN CABBAGE ROLLS

"This is a very old recipe," says Frances Koler. "I'm of Slovenian descent. We Slovaks call these 'hulupi.' The Croatians call them 'sarma.' We make pan after pan of them for weddings. They're best made a day before you want them, then reheated."

MAKES 30 OR MORE CABBAGE ROLLS

1 large head cabbage, about 3 pounds
½ cup vinegar
1 pound coarsely ground ham
2 pounds lean ground beef
1½ cups raw rice
3 eggs, beaten

1 teaspoon freshly ground black pepper
3 garlic cloves, finely chopped
2 tablespoons finely chopped onion
12 cups sauerkraut, well rinsed and drained
5 cups tomato juice

1. To soften the cabbage leaves, cut out the core and place the head in a large pot with the vinegar; cover with water. Bring to the boil and cook, uncovered, until a leaf breaks off easily, 10 to 15 minutes. (The vinegar makes the leaves firm yet pliable.) Drain the cabbage in a colander and run cold water over it. When cool enough to handle, break the leaves from the head. Cut off the hard V from the tough portion at the bottom; chop fine and set aside.

2. Mix together very well the ham, beef, rice, eggs, pepper, garlic and onion. Cut the largest leaves in half and use each half for a roll. Use the smaller leaves whole. Put two tablespoons of filling in the center of each leaf. Roll up from bottom to top, neatly tucking in the sides as you roll.

• Preheat the oven to 325 degrees.

3. Spread the bottom of a nonaluminum roasting pan with sauerkraut and the chopped cabbage. Place a layer of cabbage rolls, seam side down, on the kraut. Cover with a 1-inch layer of kraut, and another layer of cabbage rolls. Repeat the kraut-cabbage layering, ending with kraut. Pour on the tomato juice and enough water to come 1 inch above the kraut.

4. Cover and bake in the center of the oven for 3 hours. After 2 hours, add a little water, as the rice will have absorbed a lot of liquid. Next day, reheat the dish and serve.

PARK COUNTY FAIR

Livingston, Montana

FAIR DATE: LATE JULY, EARLY AUGUST, FOR FOUR DAYS

GAZETTEER: *Lying 57 miles north of Yellowstone Park, Livingston leads to one of the Park entrances. Hollywood and literary celebrities—Whoopi Goldberg, Meg Ryan, Tom Brokaw and Thomas McGuane, among many—have bought property in this spectacular big-sky, towering mountain territory (Robert Redford shot* A River Runs Through It *in and around Livingston). The bar of the refurbished 1897 Murray Hotel is the gathering place for the famous and for just folks. Fairground is on I-15.*

"We're a small rural fair," says manager Donna Goldner, "but there's sure a lot going on!" Royal titles begin at a young age here. For girls from nine to thirteen, there's a shot at the tiara for 4-H/FFA Rodeo Princess of the O-Mok-See games that raise money for FFA students. Contestants must enter at least one event in the Junior Rodeo. Kids eight and younger can ride a sheep, with the winner staying aboard for at least six seconds, which can seem like a lifetime.

Good sports enter the Calamity Jane Look-Alike Contest, named for the frontier character who dressed like a man, claimed to have been a scout for Custer and a Pony Express rider, and spent some time in Livingston. "That's always good for laughs," says Donna, "and so is the Pig Wrestling Contest. Teams have four members, coed or one sex. The pigs weigh two hundred to two hundred and twenty-five pounds for the men, a hundred seventy-five to two hundred pounds for mixed teams, a hundred and fifty pounds for women, and a hundred pounds for youths. The object is to put a pig into an upright barrel, bare hands only, and teams have one minute to pack the pig in. Actor Dennis Quaid entered this contest once."

"Our kids, Camri, Blake, and Jerid, are the fifth generation on our ranch," says Sue Isbell. "We have seven hundred and fifty acres, and we raise registered Herefords. The kids have been in rodeos all their lives, and they show cattle as 4-H'ers. We show bulls and sell them at the fair.

"It's beautiful here. And it's a hard life of physical labor; all the family must work. The kids worked as soon as they were old enough. In winter, we use our team of Belgian mares to go out and lay down hay for the cattle. We have extremes of weather. When we show our cattle at the end of January we are inside, but in those unheated barns it can be down to thirty-two below, and we still have to wash our animals. In summer, at the Billings fair, it can get up to a hundred and ten!

"At haying and branding time I do a lot of large-quantity cooking because of the numbers of people involved; and because our hours are so irregular, I cook a lot of casseroles that are easy to reheat.

"Once a year, at branding time, we have a big Rocky Mountain oyster fry. That's when the calves are castrated, and all of us save the testicles—the oysters. Breaded and deep-fried, they taste kind of like chicken-fried steak, They're not a novelty to us in this country."

"I've been in general practice here as a veterinarian for thirty years, with a large and small animal practice," says Dr. David Colmey. "Fifty percent of the practice is ranch animals, but here in this small western town I've also treated a boa constrictor with a cut nose, a circus lion, and birds both wild and domestic.

"One of the most gratifying cases, not long ago, was a young mountain lion cub for Yellowstone Park that had fallen out of a tree and broken a hind leg. A research team captured it and brought it to us. My son, Dr. Duane Colmey, and I operated on the young lion, put in some pins and screws, and it was released that day in a place where the mother was thought to be. The cub was released with a radio tracking device, and a year later it was doing well on its own.

"I'm semi-retired, and five or so years ago I became interested in cooking. I travel some and pick up recipes from other parts of the country, and two years ago I started competing at the fair. I entered a dish in the Working Mothers' Class (they let me in) and won second place. In this country we hunt, and I suppose eighty percent of our meat diet is wild game. We fish a lot, too."

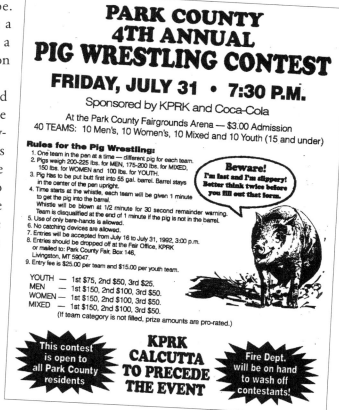

DAVID COLMEY'S BROILED ROCKY MOUNTAIN TROUT WITH PIQUANT SAUCE

"You can also grill whole trout," says David. "Place it on a double thickness of greased heavy-duty foil and cover it inside and out with Piquant Sauce. Seal the foil tightly, and place on a hot grill, five to seven minutes on each side, depending on the size of the trout."

MAKES 4 TO 6 SERVINGS

PIQUANT SAUCE:

- 1 cup chopped onion
- 2 tablespoons butter
- 2 tablespoons vegetable oil
- Juice of 1 lemon
- 1 tablespoon Worcestershire sauce
- ¼ cup honey
- 2 tablespoons ketchup
- ¼ cup dry white wine
- 1 teaspoon salt
- 1 teaspoon crumbled dry basil leaves
- 4 to 6 trout fillets, about ½ pound each

- Heat the broiler to its highest point.
1. Make the Piquant Sauce: in a large heavy skillet, cook the onion in sizzling butter and oil over moderate heat until golden. Blend in the lemon juice, Worcestershire, honey, ketchup, wine, salt and basil. Simmer over moderately low heat for 5 minutes.
2. Arrange the fillets on a greased broiler pan (or greased foil), skin side down. Spoon Piquant Sauce over each fillet and broil 5 to 10 minutes, depending on the thickness. Do not turn!

DAVID COLMEY'S HUNTERS' GOULASH

"We generally use ground elk, venison, or antelope," says David, "but I guess city folks will use beef. Serve this hearty casserole with a tossed green salad and French bread."

SERVES 6

2 tablespoons vegetable oil	1 small jar pimientos, drained and sliced
1½ pounds lean ground meat (elk, venison, antelope or beef)	1 small can mushrooms, with their liquid
¾ cup chopped onion	1 medium can stewed tomatoes, with liquid
3 stalks celery, sliced	1 can cream of mushroom soup
1 green pepper, diced	8 ounces macaroni, cooked

1. Heat the oil in a large heavy skillet or pot and cook the meat over moderate heat until it begins to brown. Stir in the onion, celery and green pepper. Add the pimientos, mushrooms, tomatoes and mushroom soup and simmer over low heat for 15 minutes. Meantime, preheat the over to 350 degrees.

2. Place the sauce in a casserole, stir in the macaroni and bake, uncovered, for about 30 minutes.

SUE ISBELL'S TOMATO BREAD

"This bread is great served with a cheese spread," says Sue.

MAKES 2 LOAVES

1½ cups tomato juice (an 11½-ounce can)	¼ cup ketchup
2 tablespoons solid vegetable shortening	1 package active dry yeast
3 tablespoons sugar	¼ cup warm water
1 teaspoon salt	6 to 7 cups all-purpose flour

1. Over moderate heat, heat the tomato juice, shortening, sugar, salt and ketchup until the shortening is melted and the sugar is dissolved. Pour into a large bowl and let cool.

2. Sprinkle the yeast over the warm water and let stand 10 minutes, or until bubbling. Add to the cooled juice mixture.

3. Gradually add 3 cups of the flour and beat it in with a wooden spoon, or use the dough hook of an electric mixer. Beat 2 to 3 minutes, or until smooth.

4. Gradually mix in 3 to 3½ cups more flour to make a soft, but not sticky, dough. Knead well on a lightly floured work surface until the dough is smooth and elastic, 10 minutes or more.

5. Gather the dough into a ball and put it in a greased bowl, turning so that all the surface is greased. Cover the bowl with plastic wrap and let rise until doubled, about 1 hour.

6. Punch the dough down, divide it into 2 parts, and let them rest and rise for 10 minutes, covered with a kitchen towel.

7. Form into loaves and put them into 2 greased 8- or 9-inch bread pans. Cover and let rise until doubled, about 30 minutes.

• Meantime, preheat the oven to 375 degrees.

8. Bake the loaves for 35 to 40 minutes, or until golden brown and hollow-sounding when thumped. Remove from pans and cool on a rack.

TWINS FALLS
COUNTY FAIR & RODEO

Filer, Idaho

FAIR DATE: FIRST WEEK IN SEPTEMBER, FOR 6 DAYS

GAZETTEER: *Filer is in south-central Idaho, about two miles from the scenic Snake River Canyon, at the junction of Highways 93 and 30. Like many fairs, the Twin Falls Fair suspended operations during World War II. However, the buildings on the fairgrounds were used—as housing for German prisoners of war.*

"Our fair began as a harvest festival in 1916," says Cindy Demoney, secretary and manager of the fair. "We're located on eighty-five acres of beautiful grass and trees. We have one man who spends four days a week just mowing and trimming.

"The rodeo is a major part of our fair. We have what's called a 'one go round' rodeo, which runs for three days. That frees the contestants who are on the rodeo circuit to go to the next event without having to spend too many days here. They want as many events as they can get. I'm very fond of the clowns, especially Rooster Kersten, with his clipped llama! (See page 262.)

"The clowns, the staff and the carnival people are among the best parts of the fair. The worst? That's easy! An opening day, Labor Day Monday, when the sewer lines clogged. The fair was jammed, and no one could flush. Plumbers don't respond too timely on holidays, so it was four hours before we got unplugged. One woman said accusingly, 'Why did you let the sewer clog on Labor Day?' As if I'd planned it!"

"Our group, the Twin Falls Magichords Chapter of SPEBSQSA, has a booth at the fair," says William Rappleye, a retired barber and hair stylist. "Those initials stand for the Society for the Preservation and Encouragement of Barber Shop Quartet Singing in America. The Society promotes four-part-harmony singing. The fair booth helps finance us, and the Society also raises money for the Institute of Logopedics, to help people with severe speech handicaps. We entertain at local functions, and often sing at the fair booth just for pleasure.

"We sell Tater Pigs. That's simply a half-pound Idaho potato with a pork link sausage baked inside. The Tater Pig takes an hour to bake, and must be served fresh and hot. We've been selling them since 1975. We bought a trailer, and good used ovens, cheap, that had been through a fire. We all participate. Some wash potatoes, some punch holes in them, some bake. Camaraderie develops. Now, you can't keep fresh food overnight, so around eleven p.m. we get pretty generous with the stuff. Once Bill Stover, who worked across the way in the Dutch oven chicken booth, came over with a bowl of chicken. We took it, filled the bowl with Tater Pigs, gave to him, and both sides had treats. We give the customers all the cheese or sour cream toppings they want, and if other concessions run out, we'll lend them some cheese. We're competitors, but so what? Working in a fair booth's a neat idea."

TATER PIGS

"We serve these with liberal toppings of butter, sour cream, hot cheese and salt and pepper," says William Rappleye. "They'll excite the most discriminatory diner."

SERVES 4

4   Idaho russet potatoes, ½ pound each
4   frozen pork link sausages

* Preheat the oven to 350 degrees.
* Scrub the potatoes well. Make a hole through each one just large enough to hold a sausage. Insert a frozen sausage in each. Bake for 1 hour or until the potatoes and sausages are cooked through.

FILER FOOD-PROCESSOR BEAN FUDGE

"We wanted to make Filer famous," says Cindy Demoney, "so we decided to host a Bean Festival every March at the Twin Falls County fairgrounds. The great state of Idaho not only produces potatoes, it's also home to one of the country's largest crops of dry edible beans. The public is invited to enter their favorite bean recipes. This fudge, a festival standby, is delicious! And you can eat it in good conscience, 'cause it's full of fiber."

MAKES 1 PAN FUDGE

1 cup pinto beans, canned and rinsed,
    or cooked and cooled
12 tablespoons (1½ sticks) butter, melted
1 tablespoon vanilla extract

1 cup cocoa powder
2 pounds confectioners' sugar
2 cups broken pecans or walnuts

1. Use a food processor to mash the pinto beans to a paste. Scrape down the bowl. With the motor running, pour in the butter. Scrape down the bowl and pulse in the vanilla.
2. Add one-third of the confectioners' sugar to the mixture, and process until incorporated. Repeat with another third; repeat with the final third.
3. With a wooden spoon, stir in the nuts. With the back of the spoon, press the fudge into a lightly greased 9 × 13 × 2-inch pan. Chill and cut into squares.

NOTE: Keep the fudge chilled; otherwise you will have a thick frosting, great for devil's food cake.

ITALIAN BEAN CUISINE

Tim Howard of the Idaho Bean Growers contributed this recipe.

SERVES 4

1½ cups chopped scallions (green onions)
2 cups sliced fresh mushrooms
1 (15-ounce) can dark red kidney beans,
    drained and rinsed
8 tablespoons (1 stick) butter

2 tablespoons red wine
8 ounces spaghetti
2 tablespoons garlic powder
1 cup grated Parmesan cheese

1. In a large heavy skillet, gently cook the scallions, mushrooms and beans in the butter and wine for 20 minutes, stirring often.
2. Meanwhile, cook the spaghetti according to the manufacturer's directions. Drain and add to the skillet. Add the garlic powder and cheese, and mix thoroughly.

ROOSTER KERSTEN, CLOWN BARREL MAN

The rodeo clown has two jobs. The first, of course, is the standard clown's job, to entertain the crowd. His second job is much more serious: to divert the attention of an angry bull away from a fallen rider.

"My barrel is an island of safety," says Rooster. "When the bull starts knocking the barrel around, I get inside. The barrels are made of real light aluminum, and they're padded with foam inside and out. Outside is to protect the bull's head, and inside is to protect my head!

"Part of my job is to hold the audience's attention by bantering with the announcer when there's a lull in the rodeo action. I fill in the spot with humor. Timing is all important to comedy, so wireless mikes have sure helped our end of the business. When you get a young announcer who's not sure of himself, you can get those little one-liner jabs in there. Comedy is all timing.

"I'll do up to fifteen county fairs a year, including Twin Falls, Idaho. I have a combined house and horse trailer, and I travel with a pony and a llama. You sure get a lot of looks when you stop to rest at roadside parks and let out that llama! Tony's his stage name. I've shaved him to look like a poodle, and in the act he's my Teenage Mutant Ninja Poodle. He's also a hunting dog. I throw out a rubber chicken, and he points just like a real dog.

"As a kid, my heroes were the clowns. I loved what they got away with, and the crowd response. My pro card from the Professional Rodeo Cowboys Association entitles me to work their shows. I think I have the perfect job, and I wouldn't trade anything for it. It's fantastic to be able to make people laugh."

ROOSTER KERSTEN'S TACO SOUP

"I cook in my trailer," says Rooster. "Friends who own the Sankey Rodeo Company gave me this recipe a while back. They helped me get my start in professional rodeo. Roberta is the brains of the outfit but we let Ike think he is. He's easier to get along with that way. This stuff is great, so be sure and try it."

SERVES 4

1 pound hamburger meat
4 cups crushed canned tomatoes
1 (16-ounce) can pinto or other beans
1 cup chopped onion
½ cup picante or taco seasoning, or to taste

Garnishes: tortilla chips, grated
    cheddar or longhorn cheese,
    chopped fresh tomatoes, black
    olives, chopped onion, sour cream

1. In a large heavy pot, cook the hamburger and the onion together over moderate heat until the meat is brown, stirring often. Spoon off excess grease, if desired. Add the tomatoes, beans and seasoning. Simmer, partially covered, for 45 minutes.
2. Spoon into 4 large bowls and let each person top with the garnishes.

CALGARY EXHIBITION & STAMPEDE

Calgary, Alberta, Canada

FAIR DATE: EARLY JULY, FOR 10 DAYS

GAZETTEER: *Cosmopolitan Calgary is in scenic western Canada, sixty miles from the Rockies, 105 miles from Lake Louise, and 73 miles from Banff. The Stampede address is 1410 Olympic Way S.E., within walking distance of downtown hotels.*

"The Stampede was started by a trick roper named Guy Weadick," says manager Brian Ratcliffe. "That was in 1912, and Weadick vowed to 'make Buffalo Bill's Wild West Extravaganza look like a sideshow.' Today, Weadick's dream venture proudly calls itself the 'Greatest Outdoor Show on Earth.'"

The Stampede kicks off a city-wide celebration. "Bands play, chuck wagons park in front of hotels, giving out free pancakes, and the business people dress Western," says Ratcliffe. "Calgary really makes you welcome.

"We've got the Half Million Dollar Rodeo, with calf roping, steer wrestling, saddle bronc and bareback riding, and bull riding; the five major rodeo events. We've also got other events, some zany, like Wild Cow

Chuck wagon races.

Milking: you have to lasso the cow and hold her still while your partner milks a measured amount into a bottle and runs it back to the scorekeeper. Not easy!

"Every evening we have the Chuck Wagon Races. Each wagon has a driver and four outriders, representing working cowboys. They have to load tent poles and a stove before they can start. There are thirty-six wagons, and they dash once around the track, four to a heat. On the tenth night of the fair, the top four compete for a fifty-thousand-dollar prize, winner take all.

"We hold a world-class Western art show, and auction selected pieces, like bronze statues of cowboys. We have a world blacksmith's competition, and a cattle auctioneer competition. They have to auction pens of cattle, as well as single animals. Entrants are judged on cadence, the ability to pick up bids and to keep track of them to maximize a sale. We also have gambling at our Frontier Casino.

"Of course we have great entertainment, but what I like best are the Young Canadians, one hundred and seventy-two Calgary youngsters we train throughout the year. We have the best of professional instructors. They audition in the fall, and they put on a terrific grandstand show, a wonderful singing and dancing revue."

CALGARY STAMPEDE RIBS AND STEAK TAILS

Dan Copthorne, president of the Exhibition & Stampede, contributed this recipe. "I prefer to use boneless short ribs or steak tails," he says. "Steak tails are the long trimmed portion of a T-bone or porterhouse steak."

MAKES 6 SERVINGS

2½ pounds boneless short ribs or steak tails	1 teaspoon salt
1 tablespoon vegetable oil	1 teaspoon paprika
1 large onion, chopped	1 teaspoon chili powder
1 (10-ounce) can tomato soup	Freshly ground pepper to taste
⅓ cup cider vinegar	

- Preheat the oven to 325 degrees.
1. Brown the meat in the oil over moderately high heat, turning often with tongs. Arrange in a baking pan in a single layer.
2. Mix the remaining ingredients together and pour over the meat. Cover the pan with a lid or aluminum foil. Bake for 3 hours, or until very tender.

SANPETE COUNTY FAIR

Manti, Utah

FAIR DATE: LAST FULL WEEK IN AUGUST, BEFORE LABOR DAY

GAZETTEER: *Manti is in the center of the state, 120 miles south of Salt Lake City. Mountain ranges ring the town. Sanpete County is rural, with ranching and farming. Coal mining is an industry. Hunting and fishing are popular pastimes, and ski resorts are nearby. The fairgrounds are on Highway 89.*

"The fairgrounds are directly below the Mormon Temple," says fair chairman Gary Myrup. "It's very scenic around here: mountains, and a little bit of desert, too. Livestock events draw a lot of interest here in ranch country. We were once a major sheep raising area, and sheep are still big. We've got a carnival, a demolition derby, an antique car show, and Baxter Black, the Cowboy Poet, reciting his works to the audience.

"Horses are important to us Westerners. We have two nights of rodeo, an open horse show, and a 4-H horse show. We crown a Sanpete County rodeo queen, whose title is Cowboy Sweetheart. I have six daughters. In 1992 my daughter Jamie was the Sweetheart, and little Mysti was Junior Princess. The following year Jamie passed on her crown to her sister Geri. I have exerted nothing! The girls won on their own merits, and I'm as proud as can be." (See page 268.)

"Each year for the parade we select a King Cowboy," says Kerry Deuel, who is in charge of the Parade and the Cookout Contest. "The king is an older gentleman who has been active in farming, ranching or anything agricultural. Someone who has helped young people get established. Ours is not the only parade in Sanpete County, but it sure is the biggest!"

TWO MARINADES FOR GRILLED LAMB FROM KERRY AND ANN DEUEL

"Both my mother and grandmother were excellent cooks," says Kerry, "and I guess I got my love of good food from them; I weigh a mere three hundred pounds! We're very fond of lamb. The Wool Growers Association, a state-wide group of sheepmen, promotes the meat as well as the wool. We do a lot with these folks."

COUNTRY STYLE LAMB MARINADE

SERVES 10

1½ cups vegetable oil
¾ cup soy sauce
¼ cup Worcestershire sauce
⅓ cup orange juice
3 tablespoons dry mustard
1 tablespoon coarsely ground black pepper

2 teaspoons salt
2 tablespoons chopped fresh parsley
2 cloves garlic, minced
4 pounds boneless leg of lamb, cut
    into 2-inch cubes

1. In a large bowl, mix all the ingredients (except the lamb) well. Place the cubed lamb in the marinade, refrigerate, and let the lamb marinate for 24 hours.
2. Pat the lamb cubes dry. Thread them on skewers and grill over white-hot coals.

BARBECUED LAMB MARINADE

SERVES 10

1 tablespoon vegetable oil
½ cup water
2 tablespoons red wine vinegar
1 tablespoon Worcestershire sauce
¼ cup fresh lemon juice
1 teaspoon dry mustard
    Dash of hot sauce

¼ teaspoon paprika
1 clove garlic, minced
1 large onion, grated
⅓ cup ketchup
½ teaspoon salt
4 pounds boneless leg of lamb, cut
    into 2-inch cubes

1. In a large bowl mix all the ingredients (except the lamb) well. Place the lamb in the marinade, turning to coat each piece. Cover and refrigerate for 24 hours.
2. Thread the lamb cubes on skewers and grill over white-hot coals.

SENTIMENTAL VIEWS OF SANPETE

At the fair, the contestants for Cowboy Sweetheart and Junior Princess are judged on a number of qualities; horsemanship, general appearance, poise and personality, and an original composition in verse called "A Sentimental View of Sanpete." These excerpts are from the winning Myrup girls:

Jamie, Geri and Mysti Myrup.

MYSTI (Age 8)

"How do I view Sanpete? I view Sanpete as a home, where I'm being raised, where I go to school with friends, and where I enjoy the greatest way of life.

Sanpete County's where I live,
It's where I hang my hat.
Pretty mountains, lovely lakes,
Sanpete County's where it's at.

We make believe we're Calamity Jane
Belle Starr, or Sluefoot Sue.
We can be a Rodeo Queen, a clown
Or a bull rider, too.

I can be a Princess
And have my own Prince Charming,
I can be a housewife,
My husband could be farming.

Here in Sanpete you can be
The person you wish to choose.
Just living here among your friends
There's just no way to lose."

GERI (Age 13)

"My heritage comes from ancestors who came to the West as early Mormon pioneers. I am a part of all the women in the West who followed their hearts and dared to be what they were:

I have a heritage of Western gals
Who lived in years gone by.
They changed the times for a gal like me,
So I could give anything a try.

I'm rough and tough, with an iron will,
On that my folks agree,
But all is softened with a smile so shy,
And a wit that's part of me.

My Western heritage is all wrapped up
And standing on display.
I'm thanking them all for what they were,
So I can be what I am today.

JAMIE (Age 16)

"Cowboys, cowboys, everywhere, and I am finally *sixteen! I am now allowed to drive, to date, be out later on weekends, and be on my own at the Sanpete County Fair Rodeo! The American flag waving in the evening breeze stirs your patriotism, but the real and only reason I go to the Sanpete County Rodeo is...the cowboys!*

YES!

Ever been to a rodeo and
Watched the cowboys walk?
Seen the twinkle in their eyes
As you listen to 'em talk?

Just by watchin' you kin tell
Which event that feller's in.
Bulls or broncs or timed event,
Or if he's gonna win.

I really have no preference
To the cowboy short or tall.
Scrawny, muscled, fat or bald,
This cowgirl loves 'em all!

MARICOPA COUNTY FAIR

Phoenix, Arizona

FAIR DATE: THIRD AND FOURTH WEEKS IN MARCH, FOR 11 DAYS

GAZETTEER: *The Maricopa County Fair is held at the Arizona State Fairgrounds, at 19th Avenue and McDowell.*

"The upside of being tenants on the state fairgrounds is that we have the use of a beautiful facility," says fair manager Dave Collins, "but the downside is that we can't rent out the facility to make ends meet if bad weather cuts our attendance. We did hold the fair under tents a few years back, but a severe storm blew the tents down, and some of the animals were injured. It's much better to be here.

"Both 4-H and FFA play important roles in our fair. I grew up in 4-H; I showed sheep, did electrical projects, and did cooking, too. Some kids who compete are the third generation of 4-H and FFA families. There's a lot of agriculture here, because the Maricopa Indians brought in irrigation. We're one of the largest carrot producers in the nation, and we have citrus and cotton, too. At certain times of the year the only lettuce you'll see is from Maricopa and our neighbor, Yuma County.

"We stress the environment. At the end of the fair we have people pick up the animal bedding to use as compost; with all the animals we show, that's a lot of manure! And we recycle a ton and a half of cardboard and a mountain of polystyrene cups. We're proud that sixty-four percent of our solid waste does not go to the landfill. That's considered exceptionally successful recycling.

"The Raptor Educational Program, using bald eagles, falcons and other raptors, emphasizes these birds' place in the food chain. All of these birds have been harmed in some way by humans—shot and blinded, or maimed—and cannot be released into the wild.

"Five times a day we have armadillo races. The Texas Armadillo Association uses the races to teach kids to avoid drugs. The armadillo's only defense is his armor, and for kids, moral values can be armor. Here's how the race works: The showman picks three kids and three adults from the audience to form three teams. A team of kid and adult gets down on their hand and knees, and they blow on an armadillo's tail. This is supposed to encourage him to run. Sometimes it works! Often they just run in circles, or sit down on their rear ends.

Armadillo races.

"We are a family fair, and we want to keep the admission at four dollars. I like to compare it to the cost of a movie, which is a lot more, and you're only there for two hours. Here you can have a day of fun for less money."

RECIPES FOR THE MARICOPA COOK BOOK

Lots of fine cooks contributed to this little book put out by the fair board, including some famous folks. Joe Garagiola gave his Meatballs and Sauce. Below are recipes from former Governor Rose Mofford, a Maricopa Fair booster, and singer Glen Campbell.

THE GOVERNOR'S CHOCOLATE CINNAMON FUDGE CAKE WITH FUDGE PECAN FROSTING

"Speedy to put together," says Governor Rose Mofford.

MAKES 1 CAKE

THE CAKE:

2 cups all-purpose flour	1 cup water
1 cup granulated sugar	1 teaspoon baking soda
1 cup packed brown sugar	1 teaspoon ground cinnamon
8 tablespoons (1 stick) butter	¼ teaspoon salt
½ cup Crisco	½ cup buttermilk
¼ cup cocoa powder	2 eggs, slightly beaten

THE FROSTING:

6 tablespoons milk	1 box confectioners' sugar (3½ to 4 cups)
¼ cup cocoa powder	1 teaspoon vanilla extract
8 tablespoons (1 stick) butter	1 cup chopped pecans

- Preheat the oven to 375 degrees.
1. Whisk the flour and sugars together in a large bowl. Place the butter, Crisco, cocoa and water in a saucepan and bring the mixture to a rapid boil. Pour it over the flour mixture and stir well.
2. Add the baking soda, cinnamon, salt, buttermilk and eggs. Blend rapidly, and pour into a 9 × 13-inch baking pan. Bake for 30 minutes. Do no overbake; a soft velvety texture is special to this cake.
3. As the cake bakes, make the frosting: Place the milk, cocoa and butter in a saucepan, heat and blend with a whisk. Off the heat add the sugar, vanilla and pecans. Spread the hot frosting on the hot cake.

GLEN CAMPBELL'S CHILI CON CARNE

This controversial dish is not really Mexican in origin, but probably was invented in Texas where chuck wagon cooks fed it to hungry cowboys on the range. There is as much disagreement over the making of a "bowl of red" as there is over the tactics of a Civil War battle. Some chili buffs say angrily that chili should be pure, that is, cooked without beans. Others disagree on including cumin seed, beef suet, tomatoes, garlic and onions.

SERVES 8

3 pounds beef chuck, coarse grind	1 quart water
2 or 3 medium-sized onions	Salt
1 bell pepper	Ground black pepper
1 or 2 cloves garlic	2 to 3 tablespoons chile powder
½ teaspoon oregano	2 (15- to 17-ounce) cans pinto or
¼ teaspoon cumin seed	kidney beans
2 small cans tomato paste	

1. Brown the chuck in an iron kettle. (If you don't have an iron kettle, you are not civilized. Go and get one.) Chop the onions and bell pepper and add to the browned meat. Crush or mince the garlic and throw it in the pot, then add the oregano and cumin seed. (You can get cumin seed in the supermarkets nowadays.)

2. Now add the tomato paste. If you prefer canned tomatoes or fresh tomatoes, put them through a colander. Add about a quart of water. Salt liberally and grind in some black pepper, and for a start, 2 or 3 tablespoons of chili powder. (Some of us use chili pods, but chili powder is just as good.)

3. Simmer for an hour and a half or longer, then add your beans. Pinto beans are best, but if they are not available, canned kidney beans will do. Simmer another half hour. Throughout the cooking, do some tasting from time to time, and, as the *Gourmet Cookbook* puts it, "correct seasoning." When you have got it right, let it set for several hours. Later you may heat up as much as you want and put the remainder in the refrigerator. It will taste better the second day, still better the third and absolutely superb the fourth. You can't even begin to imagine the delights in store for you one week later.

YAVAPAI COUNTY FAIR

Prescott, Arizona

FAIR DATE: THIRD THURSDAY AFTER LABOR DAY, 4 DAYS

GAZETTEER: *Prescott, the "Mile Hi" city, is ninety-six miles north of Phoenix. This resort town's population triples in the summer. The Prescott Frontier Days celebration, five days that take in the Fourth of July weekend, has the World's Oldest Rodeo, begun in 1888. For the fairgrounds, take Miller Road out of Prescott; from Phoenix, take I-17 to the Cordes Junction/Prescott exit.*

"We have an old fair," says manager Dora Kittridge. "Our origins date back to 1913, to the Northern Arizona fair, when the movie cowboy Tom Mix was the program chairman. He raised money, and staged a Wild West show with local cowboys.

"We like to honor our youth and our senior population, so our fair holds Kids' Day and Senior Citizens Day, with free admission for those groups. Lots of youngsters show rabbits, and we have a lecture on 'The What and How of Showmanship.' as well as 'Rabbit Talk for Adults and Kids,' dealing with rabbits as show animals and pets, and also as food.

"Surveys seem to indicate that few people are happy in their jobs, but I've found that fair managers are exceptions. My job is rewarding. Because I enjoy variety and changing activities each day, punching a time clock or working a nine-to-five job would be impossible for me.

"We have a real variety of food at the fair, and I don't let vendors duplicate; the competition wouldn't do them any good. I do make an exception for Indian fry bread. It's so popular that two booths are kept busy."

Square dancing in the old green building.

YAVAPAI INDIAN FRY BREAD

Fairgoers like the bread smeared with honey.

MAKES FOUR 6-INCH BREADS

1½ cups all-purpose flour
2 teaspoons baking powder
½ teaspoon salt
½ cup lukewarm water

2 tablespoons solid vegetable shortening, cut into ½-inch bits
Vegetable oil for deep frying

1. Sift the flour, baking powder and salt into a deep bowl. Add the shortening bits and, with your fingers, rub the flour and fat together until the mixture resembles coarse meal.
2. Pour in the lukewarm water all at once. Gather the mixture into a ball with your hands.
3. On a lightly floured surface, knead the dough with the heel of your hands for about 5 minutes, until the dough is smooth, elastic and shiny. Gather it into a ball, drape a towel over it, and let it rest for 15 minutes.
4. Meanwhile, pour 2 inches of oil into a large heavy saucepan and slowly heat it to a temperature of 400 degrees on a deep-fry thermometer, or until a bit of dough dropped in sizzles and rises.
5. Divide the dough into 4 small balls. Roll each one out into a circle about 6 inches in diameter and 1/4 inch thick. Cut three 3-inch parallel slits in the center of each round.
6. Fry the rounds, one at a time, for about 2 minutes on each side, turning once with tongs. Serve warm, with honey.

EUREKA COUNTY FAIR

Eureka, Nevada

FAIR DATE: FIRST OR SECOND WEEK OF AUGUST, FOR 3 DAYS

GAZETTEER: *Ranching and hay ranching country, formerly gold and silver mining country. For the fair, take Highway 50 from Reno.*

"*Life* magazine once featured the road from Fallon, outside Reno, to Eureka as 'the loneliest road in America,'" says Ethel Buffington. "You can pick up a map in Reno or Fallon, have it stamped at gas stations along the way, and get a free T-shirt here in Eureka, which calls itself 'the Loneliest Town.' A lot of people don't like that. They think it should be called 'the Friendliest Town.'

"A few years back, a survey done at Stanford University said that the poorest people per capita in the United States were in Eureka County. People were up in arms! So they gathered up food, made a CARE package, and sent to the man who did the survey. We have low wages, but it's hard to find the needy. When we got together stuff to give away at Christmas, we had trouble finding recipients. We're pretty sparsely populated, and you'd have a tough time finding a fair smaller than ours!

"Because of all the hay ranching in the Diamond Valley, there's a lot of competition at the fair for the Chrome Pitchfork in the Quality Alfalfa Hay Contest. Judges inspect the alfalfa visually for color, odor, leaf-to-stem ratio, and the amount of foreign matter like weeds or mold in the stack. Then there's chemical analysis to find the total digestible nutrients.

"Eureka was a gold-mining town, and a highlight of the 1992 fair was the return of the restored 1885 Eureka and Palisade engine. The train once connected to Elko, then to Reno, and carried gold, silver, cattle and people. In the mining days, there were thousands of people around here. When the engine came back, the Lions Club donated wood for the engine to burn, and the fire department brought up water twice a day. After the fair, the engine went back to Reno, and then on a tour of Europe.

The 1885 Eureka and Palisade engine.

"We have a lot of fun. For one dollar, you can throw a pie at the sheriff, and we have a cow chip throwing contest, too. There's a raffle, and you can win stuff like an oil, lube and filter job from the Ford dealer, sacks of dog food, and an electric can opener. Once I donated a crystal vase to the raffle.

"My husband, Aaron, and I run the barbecue for the Saturday night dance at the fair. For ten dollars, you get all you can eat and you dance to a live country-rock band. The whole town shows up.

"We dig a six-foot pit near our home, gather old cedar stumps and two pickup truckloads of dead hardwood to burn. We don't use any wood with varnish, creosote or paint. The wood in the pit burns till two in the morning. All the meat is donated—pork, turkey, beef and lamb. When the coals have burned down, we bury the meat, and we also cook cowboy beans, with garlic, bacon and tomatoes, underground. What we are creating is a giant pressure cooker. We've used those coals again two weeks later!"

BUFFINGTON'S BARBECUE

"Of course we barbecue six or seven hundred pounds of meat for the Eureka Fair, but I've tried to simplify what we do," says Ethel. "The pit becomes extremely hot, so use caution."

SERVES 15 OR MORE

THE DRY RUB:

2 tablespoons garlic powder	10 pounds beef chuck or shoulder roast
2 tablespoons salt	1 cup good barbecue sauce
1 tablespoon coarsely ground black pepper	3 ribs celery
3 tablespoons ground oregano, crumbled	

1. Dig a hole at least four feet deep by four feet round. Start a fire in dry ground about 8 hours before you put in the meat. (Use a good hard wood; you will need a great deal of it.) After the wood has burned its allotted time the pit should be about two-thirds full, with 2 1/2 feet of coals.
2. Lay the meat on a piece of heavy-duty aluminum foil large enough to wrap it. Rub the dry rub into the meat well, cover with the barbecue sauce, and lay on the celery ribs. Wrap the meat securely in the foil.
3. Wrap the meat package in burlap, and secure it with wire. Submerge the meat in water to wet the burlap thoroughly before lowering into the pit.
4. Lightly spray the coals with water. Stand back, as steam will rise. Place a sheet of metal (we use old rusty tin) over the coals. Lay on the meat, then cover with another sheet of metal. Shovel 1 foot to 1 1/2 feet of dirt over the meat. Leave to cook for 8 hours. (If the ground is damp or wet, leave the meat in it for 16 hours.)

SWISS CABBAGE PIE

"When he was a kid on the ranch, my husband's grandmother took this down to the hayfield around two p.m. as an afternoon snack with coffee," says Ethel.

"I collect cookbooks, and when we lived on a ranch I'd find a great-sounding recipe and realize I didn't have most of the ingredients called for. We were seventy-five miles from one town, and a hundred and fourteen miles from another. You learn to substitute. I've created a few new recipes out of necessity."

SERVES 6 TO 8

1 loaf unbaked bread (can use frozen purchased dough)	¼ teaspoon pepper
	3 eggs, beaten
1 pound bacon	1½ cups cream or half-and-half
8 cups shredded cabbage	½ cup grated Swiss cheese (optional)
½ teaspoon salt	

• Preheat the oven to 350 degrees.

1. Roll out the bread dough about 1/4 inch thick and place it in a 9-inch pie pan. Crimp the edges, as you would a pie.

2. Cut the bacon into small pieces and brown them in a large heavy pot. Pour off all but 1 tablespoon of the grease. Stir in the shredded cabbage and cook over moderate heat, stirring, for 1 minute.

3. Over low heat, add the salt and pepper, eggs, cream and the cheese, if you are using it. Cook for 3 to 4 minutes, stirring. Put the mixture in the crust and bake for 30 minutes, or until the filling is set and the edges are golden brown.

EIGHT

THE WEST COAST

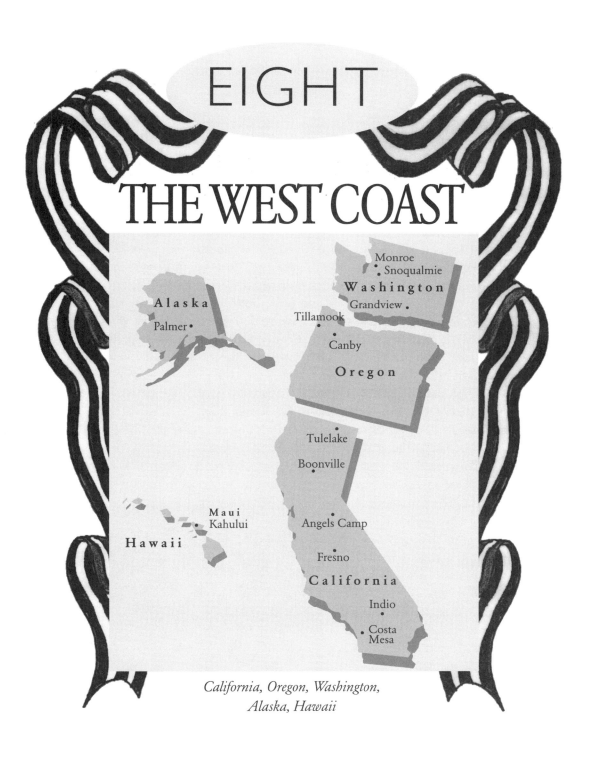

California, Oregon, Washington,
Alaska, Hawaii

RIVERSIDE COUNTY'S NATIONAL DATE FESTIVAL

Indio, California

FAIR DATE: FEBRUARY, BEGINNING THE FRIDAY BEFORE PRESIDENTS' DAY, FOR 10 DAYS

GAZETTEER: *The festival is held at the Desert ExpoCentre, 10 minutes south of I-10 on Highway 111. (The distance from Los Angeles, via I-10, is about 140 miles.) Indio is 16 miles southeast of Palm Springs.*

Want a date? Want it in an exotic Arabian Nights setting? Then come to Indio for a fair and festival like no other.

"Dates were planted in quantity here in Riverside County in the Twenties," says Bill Arballo, media coordinator for the festival. "The climate is the same as Arabia's, so it was a natural. But the early festivals had a Western theme, inspired by movie cowboys like Gene Autry and Hopalong Cassidy who had winter homes here."

Now enters a man named Bob Fullenwider, who became manager of the festival in 1947. Looking in the archives, he found that at the first festival in 1921, seven ladies from a women's club dressed up in costumes with filmy veils and made a sensation.

"That was all Fullenwider needed," says Arballo. "He adopted an Arabian Nights theme, building a Moorish-inspired wall along Highway 111. He painted quonset huts left over from the war with moons and stars, and crowned a Queen Scheherezade. And he inaugurated camel and ostrich races.

"Those races are a big hit. Camel jockeys are professionals who know how to handle these animals. As for the ostriches, you can't depend on them to go where you want. The ostriches pull little two-wheeled buggies, like you'd see in a harness race. The driver has a small broom, and uses it to guide his bird by covering one eye or the other. Sometimes an ostrich will spot a tasty-looking hat in the audience, dash over to the stands, and try to eat it.

Neck and neck: the Camel Races.

"We have a Moorish stage, and a nightly '1001 Nights' pageant. The pageant involves about a hundred people; by day they're doctors, lawyers and merchants, by night they're make-believe sheiks and harem beauties. Rehearsals start in November. You can be sure that despite her many trials the princess will be spared at the end of the fantasy; it may take some doing, but she's always been spared.

"If you come to the festival in a costume, you can get in free; it just has to be vaguely Arabian. Costumes add to the atmosphere. There's plenty to see and do! We've got free-style arm wrestling, ladies' classes too; a terrific carnival; and a wonderful model railroad with its own carnival and working Ferris wheel, all to scale. There's a big Mexican village; the dominant culture here is Hispanic, because of the agriculture. And, of course, date displays, and date, citrus and Mexican cooking contests."

WHOLESOME DATE BARS

Patricia Duff, of Desert Hot Springs, was a winner at the Date Festival with these bars.

MAKES 24 BARS

3 cups chopped dates

1 cup water

8 tablespoons (1 stick) butter, at room temperature

¾ cup packed brown sugar

1 teaspoon vanilla extract

2 cups quick or regular rolled oats

2 cups whole wheat flour

½ teaspoon baking soda

• Preheat the oven to 325 degrees.

1. Simmer the dates and water together until they form a thick paste, about 15 minutes. Set aside.

2. In a large bowl, beat together the butter and brown sugar until fluffy; beat in the vanilla. Add the rolled oats, mix well. Add the flour and baking soda and combine well, forming a crumb mixture.

3. Lightly grease a 13 × 9 × 2-inch baking pan. Place half the crumb mixture in the pan and pat down firmly. Spread with the date mixture. Spread on the remaining crumb mixture, and pat firmly. Bake for 25 to 30 minutes, until lightly browned. Cool on a cake rack, and cut into bars.

TANK TROOP DATE SHAKES

During World War II, General Patton's tankers trained in the desert east of Coachella Valley, where 95 percent of American dates are harvested. Local residents could make a pretty penny selling thirst quenchers to the servicemen. Enterprising date grower Russell C. Nicoll invented the date milkshake, an instant hit with the thirsty soldiers. He opened a roadside stand and put his daughter in charge of it. The Valerie Jean Date Shop achieved real fame; it can be found on most published California maps. Today there are a number of roadside date stands in the area. Nicoll was also the first to market dates in cellophane packages. Before that, they were sold in paper bags.

MAKES 1 DATE SHAKE

3 scoops vanilla ice cream   ½ cup milk
½ cup pitted dates

• Put the ice cream in a blender container and whirl until liquefied. Add dates, blend until finely chopped. Add milk, blend briefly, and serve.

Date Festival Princess, with camel.

ORANGE COUNTY FAIR

Costa Mesa, California

FAIR DATE: MID-JULY, FOR 17 DAYS

GAZETTEER: *Populous Orange County was once covered with groves, as its name implies, and the fair pays homage to its agricultural roots. The fairgrounds are at 88 Fair Drive in Costa Mesa, off exit 405 on the San Diego Freeway.*

Fair beauties, 1920s.

"In 1992 we celebrated our hundredth anniversary," says promotions director Diane Sorenson, "and the theme was 'One Hundred Years in the Making.' We buried a time capsule containing memorabilia from past fairs. County residents who were a hundred years old came on opening day for a party, and Willard Scott came to the party too, for the *Today* show!

"And as 1993 was the hundredth birthday of the Ferris wheel, we felt it was appropriate to honor it. Now, this is California, and we've got the Ferris wheel-riding champion of the world, Jeff Block. His record was thirty-seven continuous days on a wheel, and his goal was to break it, with thirty-eight days on our wheel. He did. We had an 1949 antique wheel, set up in the main mall. We also had an extensive historical display, and we called all this hoopla 'It's the Wheel Thing.'

"In addition to the Ferris wheel celebration, our other 1993 theme was 'We're Having Bushels of Fun,' saluting Orange County's vegetable industry. On opening day, you could see a giant stalk of celery, a jicama bulb, an artichoke and a pair of tomatoes floating down from the heavens; the 'vegetables' were costumed skydivers. For the Corn Derby, we asked for corn kernels from every state, and planted them here in April; the highest stalk by July ninth got the Golden Ear award. The winner was from South Carolina, topping off at eight feet, eleven inches.

"The Couch Potato contest drew two of the most 'creatively inactive' people in Orange County. Each one had to set up a living room vignette in the Harvest Tent and decorate it. We provided the couches and junk food. It was worth it to lounge around for seventeen days; the winner, Al Harris, of Irvine, got fifteen hundred dollars and a lifetime membership to a fitness club."

RECIPES FROM THE ORANGE COUNTY FAIR
CENTENNIAL COOKBOOK, 1992

For its hundredth-year celebration, the fair issued a small cookbook full of good recipes from local residents. Many of them (recipes with "a-peel") honor Orange County's past as a center of California's citrus industry.

ORANGE COUNTY'S BEST SALAD

Paula Vogelsohn of Los Alamitos says, "I always use oranges in salad when tomatoes are high. I had never tasted an avocado before moving to California thirty years ago."

SERVES 4 TO 6

1 head iceberg lettuce, torn into
   serving pieces
1 bunch watercress (or 1 bunch cilantro)
2 oranges, peeled and sliced
1 ripe avocado, peeled and sliced
1 (6-ounce) jar marinated artichoke hearts

2 tablespoons vinegar
1 tablespoon Dijon mustard
2 scallions, thinly sliced (or 2 tablespoons
   chopped red onion)
2 tablespoons toasted sesame seeds

1. Just before serving, toss the lettuce and watercress together in a serving bowl. Arrange the orange and avocado slices over the greens. Put the artichokes on top, and save the marinade in the jar.
2. Add the vinegar, mustard and scallions to the jar, tighten the lid and shake. Pour the dressing over the salad and sprinkle on the sesame seeds.

ANANIA'S MINESTRA DI PANE (BREAD SOUP)

Marilyn Anania of Huntington Beach says, "This soup is very filling and could be considered a whole meal. Serve with a fine white wine or mineral water."

SERVES 12

2 tablespoons olive oil
¼ cup vegetable oil
1 large onion, chopped
1 medium head cabbage, shredded
½ cup chopped celery
½ cup chopped carrots
2 small potatoes, cubed
½ cup tomato sauce
   Water to cover the vegetables

1 teaspoon salt
½ teaspoon freshly ground white pepper
1 cup freshly cooked or drained
   canned white Great Northern beans
1 to 2 loaves day-old Italian bread,
   cut into thin slices
   Extra-virgin olive oil, for drizzling
1½ to 2 cups freshly grated Parmesan
   or Romano cheese

1. In a large, preferably nonstick, pot, heat the two oils and cook the vegetables and the tomato sauce, stirring often, for 15 minutes over low heat.
2. Cover with water and add the salt and pepper. Bring to the boil, lower the heat to moderate, cover and cook 1 to 1¹/2 hours, stirring from time to time.
3. Add the beans and simmer 15 minutes to blend flavors.
4. Thirty minutes before serving, line a very large bowl or tureen with bread slices. Ladle on some of the vegetable mixture. Keep making layers of bread and vegetables, ending with vegetables. Let stand 15 minutes, so the bread slices can absorb the liquid.
5. Ladle into individual soup bowls, and pass olive oil and grated cheese to each person.

HAPPY HUNDREDTH BIRTHDAY TO THE FERRIS WHEEL

Think "fair" and what comes first to mind? The Ferris wheel, of course. It's been delighting fairgoers for a century.

Twenty-seven million people visited the great World's Columbian Exposition in Chicago in 1893, and most of them must have gaped at the enormous churning wheel built by engineer George Washington Gale Ferris. A goodly number rode the behemoth. It was, and remains, the largest Ferris wheel ever built. The axle alone weighed fifty-six tons, and remains the largest piece of steel ever forged. The highest point of the colossus was two hundred and sixty-five feet off the ground; the diameter of the wheel was two hundred and fifty feet, and the circumference eight hundred and twenty-five feet. A thousand-horsepower engine powered the wheel, which revolved the first time Mr. Ferris turned it on. The ride carried thirty-six cars, each capable of holding forty to sixty riders. At capacity, two thousand one hundred and fifty people could ride Ferris' creation at one go; a single turn of the wheel took twenty minutes.

The Ferris wheel next appeared at the St. Louis Exposition in 1904. It was then trundled back to Chicago and dismantled; it was just too big and expensive to maintain. This first wheel met a patriotic end, though; its metal was used to build the naval vessel U.S.S. *Illinois*, which served in World War I.

MOROCCAN STRING BEANS AND CILANTRO

Karen Green of Irvine was a 1985 Orange County Fair demonstrator. She has written a number of cookbooks, written for magazines, and appeared on television.

MAKES 4 SERVINGS

1　pound string beans, ends trimmed

SPICY DRESSING :

4　teaspoons tomato paste	½　teaspoon paprika
2　tablespoons lemon juice or vinegar	1　tablespoon olive or other vegetable oil
1　large clove garlic, minced	Salt and pepper to taste
1　teaspoon ground hot red pepper	3　tablespoons chopped cilantro
1　teaspoon ground cumin	

1. Slice the string beans into thin strips about 2 inches long. Boil in water to cover for 4 to 5 minutes, until barely tender. Drain and rinse immediately with cold water, to stop cooking and set color. Place the beans in a bowl.

2. Make the dressing: Put the tomato paste in a small bowl and mix with lemon juice. Add the garlic, red pepper, cumin and paprika, stirring to blend well. Mix in the olive oil. Season to taste with salt and pepper. Pour over the beans and toss well. Let sit at room temperature for several hours to allow flavors to mingle. Just before serving, chop the cilantro and toss it in. (The salad can be made a day in advance, refrigerated, and allowed to come to room temperature before serving.)

THE BIG FRESNO FAIR
Fresno, California

FAIR DATE: EARLY OCTOBER, FOR 17 DAYS

GAZETTEER: *Fresno is in the San Joaquin Valley, the most productive agricultural area in the country. Nearly two hundred and fifty crop varieties are grown here, including table and raisin grapes, citrus, melons, peaches and apricots. The Fresno Fairgrounds are off Highways 41 and 180, at 1121 Chance Avenue.*

"We have the largest agricultural display in the nation; more than twenty-three hundred different categories of dried and fresh fruits, vegetables, hay, crops and nuts," says Maggie Walen, in charge of public relations for the fair. "We've got it all, because we have a three-hundred-and-sixty-five-day growing season. Over twenty-five communities around Fresno participate; we're said to be the fastest-growing city in the United States. We have huge 4-H and FFA participation, and school tours of

The Midway.

the fair, with lesson plans for the children to learn about agriculture and livestock. There is horse racing every day, big-time entertainment every night. We're starting a large-scale farmers' market in one of our buildings. The fairgrounds are used throughout the year; in May we have Fiesta Days to honor the Hispanic community; Fresno is 36 percent Hispanic."

"My folks are from Durango, in Mexico," says Tommie Arenas. "My dad worked for the railroad, and worked at farming. He irrigated grapes and cultivated cotton. We lived ten miles from Fresno, surrounded by fig orchards. The figs and cotton are gone now; the city has spread out.

"As a kid, I picked cotton. We got out in the fields at four p.m., after school. We'd work for two or three hours, wearing gloves. We picked figs, too; they were spread on the ground to dry. Us kids got to keep a little of the money we earned. It was a hard life, but we didn't mind it. We wanted to help our parents."

"My grandparents were wheat farmers, who came to the Valley in 1851," says Walter J. Gilgert, a retired investment broker and commodity futures trader, who also wrote articles on agricultural commodities for a farming newspaper. "People followed the railroad, as it went from Sacramento and Stockton to Bakersfield. People came from various countries and staked out their areas. There was a Danish colony, and a Swedish colony; some settlers even bought land parcels before they came to this country. Armenians are heavy in the Valley—they came after the Turkish massacres in World War I. Fresno was organized around 1885. This was a desert before irrigation; the early people diverted water from the San Joaquin and Kings rivers. There are irrigation companies all over the area.

"The Valley is cut in half by Highway 93; on the eastern side are citrus groves, on the western side are grapes and fruit trees. The two sides have different soil, and there's more rain in the east. Farmworkers on fruits and crops are about eighty percent Mexican-American. We've had fifty-five thousand Asian refugees, including Laotians and Hmong. Many have bought land, or are leasing, and are doing well. Besides the crops like strawberries and string beans, they're growing Asian crops; you can find bitter melon and bok choy at the Farmers' Market."

TOMMIE ARENAS' CHILI BEANS

"My husband is a member of the Optimists Club, and they have a picnic for a large number of children every year," says Tommie. "I prepare my chili beans for the treat. My original recipe is for ten pounds of beans, which feeds about a hundred and fifty. This version is for a family."

SERVES 6 TO 8

4 cups pinto beans, rinsed and picked over, but not soaked	1 carrot, cut up
2½ quarts water	3 to 4 tablespoons Gebhardt's Chili Powder
½ pound ground round steak	1 (8-ounce) can tomato juice
¾ teaspoon cumin seed	2 slices bacon, cut small
2 cloves garlic, peeled	1 tablespoon salt, or to taste

1. In a heavy kettle, bring the beans and water to the boil. Lower the heat, cover and cook gently for 2 hours, checking often to see that the beans are just covered with water. If necessary, add a little boiling water. Stir from time to time.
2. Cook the ground round in a skillet until well browned. Drain off any fat.
3. In a blender or food processor, grind together the cumin seed, garlic and carrot.
4. Stir the meat, cumin seed mixture and chili powder into the beans. Stir in the tomato sauce and bacon, and add the salt. Cook 1½ hours longer, or until the beans are tender. Keep them moist.

APRICOT BUTTER

"My butters have won me prizes at the Fresno Fair, and so have my pies and jams," says Tommie Arenas. "I often have forty-eight entries, and if I win ribbons for half of them, I think I'm doing real good. This recipe will work with other fruit, too, and is great as a glaze for baked ham."

MAKES ABOUT 4 CUPS

6 cups washed, cut-up fresh apricots, pits removed

½ cup water
Sugar

1. In a heavy nonaluminum pot, put the fruit and water, and bring to the boil. (The water is to keep the fruit from scorching until it begins to give out its own juices.) Lower the heat and simmer, covered, for about twenty minutes or until the fruit is soft, stirring often.
2. Force the fruit through a food mill or coarse sieve, then through a fine sieve. For each cup of pulp, add 1/2 cup sugar. Cook gently, uncovered, for 45 minutes, stirring often.
3. Have ready hot sterilized jars and lids. Fill the jars, leaving a little headroom, and seal. Store in a cool place.

GILGERT'S RED PEPPER RELISH

"My mother, Mabel Morris Gilgert, a native Californian, was born in Monterey County in 1894," says Walter Gilgert. "This is her recipe; I've won prizes with it at the Big Fresno Fair, where I've been competing for twenty years, at least. I use the relish on meat and chicken sandwiches, and pour it over cream cheese; throw on some cocktail shrimp, and you have an appetizer, with crackers. My sister-in-law reports that my brother doesn't think it's time to eat until the red pepper relish is on the table."

MAKES ABOUT 4 CUPS

12 large red bell peppers
1½ teaspoons pickling salt

2 cups fresh lemon juice or white vinegar
3 cups granulated sugar

1. Remove the stems, seeds and mid-ribs from the peppers. Run them through a food grinder. Put them in a nonaluminum bowl and toss them with the salt. Let stand for 1 hour.

2. Drain liquid from the peppers. In a large heavy saucepan, mix the peppers with the lemon juice and sugar. Bring to a soft boil, lower the heat and simmer for about 1 hour, until the relish thickens somewhat. Low heat is important; the relish burns easily. Stir frequently. The color should be a nice, darkish red.

3. Pack the hot relish in hot sterilized half-pint jars. When cool, wipe off the jars and store them in a cool dark place. This relish keeps for over a year.

GILGERT'S ZUCCHINI RELISH

"Another of my mother's recipes," says Walter Gilgert, "and another fair winner for me. I use the relish in homemade Thousand Island salad dressing, mix it with mayonnaise as a fish condiment in lieu of tartar sauce, use it in potato salad and deviled eggs, and in a dozen other ways."

MAKES ABOUT 10 CUPS

10	cups sliced zucchini	2½	cups white vinegar
4	cups sliced yellow onion	4½	cups granulated sugar
5	tablespoons pickling salt	1	teaspoon dry mustard
1	large green pepper, stemmed, seeded and ribs removed	1	teaspoon turmeric
		1	teaspoon celery salt
1	large red pepper, stemmed, seeded and ribs removed	½	teaspoon black pepper
		1	tablespoon cornstarch

1. First Part: Put the zucchini and onions through a meat grinder. Put them in a nonaluminum bowl and stir in the salt. Cover and refrigerate overnight. Next morning drain well, put in a large colander lined with cheesecloth, place in the sink and run cold water through the mixture. Drain again. This will get rid of most of the salt.

2. Second Part: Grind the two peppers. (Two green would do, but the red gives some spots of color and adds flavor.)

3. In a large kettle, combine the vegetables and the remaining ingredients. Bring to a soft boil. Lower the heat and simmer for 30 minutes, stirring frequently to avoid scorching.

4. Place the hot relish in hot sterilized pint or half-pint jars and seal. After 24 hours, wipe off the cooled jars, if sticky, and wash the rings and replace.

CALAVERAS COUNTY FAIR & JUMPING FROG JUBILEE

Angels Camp, California

FAIR DATE: THIRD WEEKEND IN MAY, FOR 4 DAYS

GAZETTEER: *Angels Camp, an important Gold Rush mining town, is 64 miles east of Stockton, in the foothills of the Sierras. For the fairgrounds, called "Frogtown," take Highway 49 two miles south of town.*

"Underneath Angels Camp is a honeycomb of connecting mine shafts," says fair staff member Debbie Rocco. "We stand on a hollow core. Downtown has most of the original buildings, and a lot of houses still have their old mine shafts in the basement."

In the summer of 1860, humorist Mark Twain rented a small mountain cabin near Angels Camp. On a visit to town, he overheard some saloon talk about a jumping frog named Dan'l Webster, and some shenanigans concerning a contest. Back at his cabin, he wrote "The Celebrated Jumping Frog of Calaveras County," a tale that still provokes hearty chuckles.

Jump that frog!

Leaping forward to 1928, Angels Camp's citizens were casting about for a way to celebrate a job of street paving. Yes! The Frog Jump was born. (Some say that street paving was only an excuse—the real reason was to attract respectable visitors to a town known for gambling and brothels, and to force Angels Camp to clean up its act.) In 1937, the Jumping Frog Jubilee joined forces with the itinerant Calaveras County Fair, which had moved from community to community since 1880, to establish a permanent home.

Anyone can be a frog jockey. You can bring your own, taken from a pond or stream, get one through the fair's Rent-a-Frog program, and you can hire a proxy to jump your frog for you. The frog must be at least four inches from nose to rear, and no toads are accepted. (Toads are broader and drier than frogs.)

A frog is allowed three jumps. The "jockey" places the frog, and urges it on with stomps and whistles. The jump distance is measured in a straight line from the starting pad to the third point of landing, so if your frog makes its second jump sideways, that cuts down on its distance. The Guinness Book of Records championship is held by Rosie the Ribiter, who launched herself a spectacular twenty-one feet, five and three-quarters inches in 1986. There's a five-thousand-dollar prize awaiting the frog and jockey who beat Rosie's record. There is frog jumping every day at the fair, with a final jump-off on Sunday for about seventy-five qualifiers.

As a great fair should, Calaveras has a carnival and live entertainment as well as livestock and food preservation judging. It also boasts a rodeo and an arm wrestling contest, pig races and games of skill. With or without a frog, you'll have a grand time.

"I like to look after the 4-H'ers, and bring coffee for the early birds like the security guards," says Dorothea Cooper, who has ranched since 1943 in Salt Springs Valley, here in the mother lode country. "In the mid-nineteenth century, our ranch was a stopover for stagecoaches and wagons," she notes. "People changed horses and ate meals. There were thousands of fruit trees, and the place was called the Peach Orchard Ranch. By the time we bought the ranch the fruit was long gone, and the place was pretty run down. Now we have grapes, apricots, oranges, apples, crab apples, quince and pear trees, currants, strawberries, loganberries and youngberries, which are like mild blackberries."

In addition to running a five-hundred-and-seven-acre cattle ranch, Dorothea Cooper is the queen of competitors; in a recent year she brought one hundred and fourteen entries to the fair! She estimates that over the years, she has been top winner in about 70 percent of the categories she has entered. Other honors have gone to her daughters: in 1963, Geraldine was crowned Miss Calaveras, and Judy was Miss Junior Calaveras.

DOROTHEA COOPER'S YOUNGBERRY, LOGANBERRY OR BLACKBERRY JAM

"The jam is not oversweet," says Dorothea.

MAKES 6 CUPS JAM

6　cups crushed berries

3　cups sugar

1. Put the stemmed and washed crushed berries with the sugar in a nonaluminum heavy pot. Cook over moderately high heat, stirring continually, until the mixture is thick.
2. Pour into hot sterilized jars, leaving 1/4 inch headroom. Seal according to manufacturer's directions, and process in a boiling water bath for 10 minutes. Make certain the lids are sealed. Remove the bands, wipe dry, replace.

DOROTHEA'S PORK CHOPS WITH YAMS

"A nice one-dish meal that's done in under an hour," says Mrs. Cooper. "My recipe appeared in Come and Get It, *the Calaveras-Tuolumne Cattle Women's Cookbook, 1987."*

SERVES 4

4　pork chops, ¾ inch thick

1　tablespoon fat cut from chops, or vegetable oil

1　teaspoon salt

Freshly ground black pepper to taste

¼　to ½ teaspoon dried thyme, crumbled

¼　to ½ teaspoon dried marjoram, crumbled

4　medium yams, peeled and boiled until tender but firm

1　green pepper, cut into rings

2　cups canned tomatoes, or the equivalent in fresh peeled and chopped tomatoes

1. In a heavy skillet in which the chops fit comfortably, brown them on both sides in the fat. Season them with the salt, pepper and herbs. Leave the chops in the skillet.

2. Slice the yams about 1/2 inch thick (do not overcook them or they turn to mush), and cover the chops with them. Place the pepper rings over the yams, and top off with the tomatoes. Bring to the boil over high heat, reduce the heat to low, cover, and cook the chops until tender, about 45 minutes.

JOHN SNYDER'S POW WOW JERKY

John has served as a fair board member and helped develop the rodeo and junior livestock programs. This champion barbecuer takes great pride in his pungent jerky. "I usually make six pounds at a time," John notes. "It's a great snack food for kids and adults, instead of candy or cookies." The jerky will keep for a month in a sealed tin.

MAKES 3/4 POUND JERKY

2 pounds very lean ground beef, or other meat	1½ teaspoons brown sugar
⅓ cup soy sauce	1 teaspoon salt
1 teaspoon liquid smoke	1 teaspoon pepper
	1 teaspoon pureed garlic

1. Mix all the ingredients together well. Refrigerate, covered, overnight to let the flavors blend.
2. Next day, roll out the mixture between sheets of wax paper or plastic wrap to a thickness of 1/4 inch. (For guidelines, put two 1/4-inch rulers or slats on either side of the lower sheet of paper or wrap. Lay on an amount of meat and flatten with your hand. Place on the upper layer of paper or wrap, and roll with a rolling pin. The meat will be exactly 1/4 inch high.) Place the layers of meat in a 150 degree dehydrator, in convenient pieces (or place on mesh over cake racks in a 150 degree oven). Dry the meat for 7 to 8 hours. The jerky will be chewy but not completely dry. Cut into 3/4- to 1-inch strips with scissors.

MENDOCINO COUNTY FAIR AND APPLE SHOW

Boonville, California

FAIR DATE: THIRD WEEKEND IN SEPTEMBER, FRIDAY THROUGH SUNDAY

GAZETTEER: *The county's beautiful Anderson Valley, 150 miles north of San Francisco, comprises the towns of Boonville, Philo, Navarro and Yorkville. Reach the valley on Highway 128 from the south, or 253 from the east; both are off Highway 101. The rolling terrain is covered with apple orchards, vineyards, and grazing sheep and livestock to the east, redwoods to the west. The fair takes place "smack dab in the middle of Boonville." Visitors to Anderson Valley will find roadside stands, welcoming wineries, and excellent restaurants.*

If you're looking for bahl gorms, pike to Boont. Translation: If you're looking for good food, come to Boonville. Is it a foreign language? No, a home-grown one, a tongue called "Boontling" developed and spoken in Boonville between 1880 and 1920. ("Boont" is the nickname for Boonville, and "ling" comes from lingo, or slang.) It apparently began in fun, but as the language developed, speakers—or "harpers"—found they could harp Boont before the folks from Philo, and "shark" them; that is, keep them in the dark. Everyone else, too. Now, that was really fun! Contact with the outside world inevitably caused Boontling to wither. Only a few still speak it fluently, though everyone knows at least a few words, such as "buckey walter," or telephone, which appears on the signs over the public phone booths.

Lorrie Gardner, "pin-up girl."

"We're a harvest fair," says secretary/manager Margaret Hosier. "We're the youngest wine-producing area in California, and we're making a name for ourselves with Roederer Estate sparkling wine. We also have forty varieties of apples, on about eight hundred acres. A lot of them are juiced for Martinelli Sparkling Cider and Gowan's cider.

"On the night before the fair, the lugs, or boxes, of apples come in, and people spend hours polishing them and arranging them along a hundred-and-fifty-foot wall. Boxes and boxes of beautiful apples. They're judged by the box and individually, for ribbons and money prizes. At the end of the show, visitors can buy them.

"I think we're the only fair in northern California with sheepdog trials—the valley was once strictly given over to sheep. It's fascinating to watch. Each dog is given three sheep to take through a variety of obstacles and pens, with silent instructions from the handler; all hand signals. What's *not* silent is the rodeo on Saturday and the rodeo finals on Sunday."

"Some thirty years ago my mother-in-law, Edna Sanders, a retired teacher, put together a little cookbook," says Betty Sanders. "It was called *Bahl Gorms in Boont*, or *Good Food in Boonville*. It was a labor of love for her."

Here's a sampling of Boontling's culinary vocabulary:

easter	egg	dumplin' dust	flour
charl	milk	rudy nebs	water
boo	potato	zeese	coffee
doolsey	sugar	price babcock	sweet cider
doolsey boo	sweet potato	dreek	beat

EDNA SANDERS' EASY LEMON CAKE

This recipe is in Boontling. To follow, consult the vocabulary list.

MAKES 1 CAKE

2 cups doolsey	1 tablespoon baking powder
4 easters	1 cup cream
2⅔ cups dumplin' dust	1 teaspoon lemon extract
Pinch salt	

• Preheat the oven to 350 degrees.
1. Cream together the doolsey and easters until light and fluffy.
2. Sift thribs [3 times] the dumplin' dust, salt and baking powder. To this mixture add the cream and creamed mixture. Add the lemon extract and dreek 3 minutes. Pour into 2 greased and dumplin' dusted cake pans, and bake for 25 to 30 minutes. Cool on racks. Frost with your favorite frosting.

TULELAKE-BUTTE VALLEY FAIR
Tulelake, California

FAIR DATE: September, Wednesday through Saturday following Labor Day, for 5 days

GAZETTEER: *Tulelake is three miles south of the Oregon border, 28 miles south of Klamath Falls. Nearby is the Lava Beds National Monument, site of Captain Jack's Stronghold, the last major Indian battle in California, ending the Modoc War of 1872–3. The Newell Internment Camp, where 16,000 Japanese-Americans were held during World War II, is south of Tulelake. Eighty percent of the waterfowl in the Pacific flyway funnel through the Tulelake Basin. Bald eagles live here, as do game animals such as mule deer, antelope, quail and bobcats. The fairgrounds are just west of Highway 139.*

"Everything around here used to be under water," says Cindy Wright, secretary/manager of the fair. "After World War I, the lake was drained, and homesteaders' names were drawn out of a pickle jar for sites. Visitors who come to see the Japanese camp find only the remains of a few buildings. Most barracks were given to homesteaders, who remodeled them as ranch houses. We have a museum on the fairgrounds and we're trying to gather information and pictures of the housing, then and now.

"The fair started in 1949, as a junior livestock show, with a budget of seven dollars. The Rotary put on a barbecue to raise some funds; it was felt that youth shouldn't have to travel so far to other fairs. Our fairgrounds are next to the high school, and there's a football field in front of the grandstands. The first games of the season kick off the fair; Butte Valley and Tulelake playing other teams.

"For the Greased Pig Contest, the spectators form a human wall around the field. The kids are divided into weight groups, the pigs are greased with lard, and as pigs and kids dart around the field, it's up to the "wall" to keep the pigs from escaping.

"Potatoes are the number-one cash crop around here, and the fair's logo is a little potato figure, Diamond Gem. He's named for a potato variety. He kind of looks like a caricature of Bill Whitaker, a former fair manager. The artist got a bit of him! We have a potato peeling contest, to see who can get the longest peel in the least time, and a mashed potato eating contest."

"In 1909," says Mary Victorine, "sixty Czech families colonized the town of Malin, on the shores of Tule Lake, which was then undrained. They sent for their families. They anticipated that the government would someday drain the lake, but perhaps they didn't anticipate the contrary weather—it can freeze any month of the year here. A Czech town called Malin was said to grow the best horseradish in Europe, and those were the roots the colonists brought here. Horseradish can withstand frost, and you can harvest it in the fall or in the spring; you can cut off the top ten inches, irrigate, and it will grow again.

"Now we have a sprinkler system that protects potatoes from frost, but we still like frost-resistant crops. In the early Fifties, the University of California developed a superior horseradish root, derived from that old Czech rootstock. After the lake was drained, many families moved to farms on the lake bed. Today we raise about one-quarter of all the horseradish in the country right here."

ROAST PORK WITH CARAWAY SEEDS

"This is one of our favorite almost-entirely-homegrown meals," says Mary Victorine. "We like it with a side dish of sauerkraut, browned in the roasting pan after the fat is removed. To spice it up, serve it with a mixture of three-quarters mustard, one-quarter horseradish."

SERVES 4 TO 6

1 clove garlic	1 tablespoon caraway seeds
1 (4-pound) pork loin or shoulder roast	½ cup water
Salt to taste	1 medium onion, cut into quarters

· Preheat the oven to 350 degrees.

1. Cut thin slices of garlic and insert them into little cuts in the meat. Salt it lightly, press on the caraway seeds and place the roast, fat side up, in a roasting pan with the water and onion.

2. Roast for 1³/4 to 2 hours, or until the meat reaches 160 degrees on an instant-read thermometer. Let rest 15 to 20 minutes for easy carving.

MARY'S PLUM DUMPLINGS

"We like these for Sunday evening supper," says Mary Victorine. *"Leftovers can be crisped in butter for breakfast."*

MAKES 24 DESSERT DUMPLINGS

2 eggs, slightly beaten
½ cup milk
2 to 2½ cups all-purpose flour
1 teaspoon salt

24 medium fresh Italian prune plums,
washed and pitted
Melted butter
Cinnamon, for sprinkling

Granulated sugar, for sprinkling

1. Mix the eggs and milk in a bowl; stir in 2 cups of the flour and the salt. Add more flour if necessary to make a dough that is stiff enough not to be sticky when handled. Knead well, and pat the dough out on a floured surface. Cut into 24 pieces, each about the size of a walnut.

2. Shape a piece of dough into a little cup with your fingers. Wrap it around a clean plum, covering the plum completely; pinch to seal tightly. Continue with the remaining dough.

3. Let the dumplings rest 15 to 20 minutes. Drop them into 4 inches of boiling water and cook, covered, for 15 to 18 minutes. Lift out with a slotted spoon, letting the dumplings drain. It may be necessary to cook the dumplings in 2 batches.

4. To serve hot, split and pour on melted butter, and sprinkle with cinnamon and sugar.

KOLAČES (Yeast Dough Fruit-Filled Cookies)

"You could call these Czech thumbprint cookies," says Mary Victorine. "The baked kolačes freeze very well. Traditionally they are served here at weddings. On a recent Fourth of July, my daughter Bonita Fillmore and I made three thousand of them, and sold them to benefit the park in Malin, over in Oregon. "

MAKES 6 DOZEN KOLAČES

2 packages active dry yeast	½ teaspoon grated mace
¼ cup lukewarm water	1½ teaspoons salt
1 tablespoon sugar	½ teaspoon grated lemon rind
½ pound (2 sticks) butter	6 to 7 cups all-purpose flour
2 cups milk	½ cup melted butter
3 eggs	Poppy Seed Filling (see next page)
½ cup sugar	1½ cups ground nuts

1. Sprinkle the yeast over the water; add the sugar and let sit until bubbling, about 10 minutes.

2. In a heavy saucepan, heat the butter and milk over moderate heat until the milk is warm and the butter melted.

3. In a bowl, bread maker or in a heavy-duty electric mixer, beat the eggs and sugar together until the mixture is light and thick. Add the milk-butter mixture, the yeast, mace, salt and lemon rind.

4. With a wooden spoon, or in a bread mixer equipped with a dough hook, beat in the 6 cups of flour, 1 cup at a time. If the dough seems liquid, beat in some or all of the remaining flour. When the dough is too thick to mix with the spoon, or pulls away from the side of the mixer, turn it out onto a floured board. Knead 8 to 10 minutes until the dough is smooth and no longer sticky. Place in a buttered bowl, turn to coat, cover with a kitchen towel and let rise in a warm place until doubled in size, 1 to 1 1/2 hours.

5. Turn the dough out onto a floured surface and cut into 6 large pieces. Divide each piece into 12 small ones.

6. Form walnut-sized balls by rolling the dough on a surface with your palm. Place 2 inches apart on greased baking sheets. Brush each ball with melted butter. Let rise 30 minutes or until almost doubled in size.

• Preheat the oven to 400 degrees.

7. With your finger, press a nest in the center of each ball, so that the dough forms a shell 1/2 inch thick. Fill with your poppy seed filling. Top with a sprinkle of ground nuts. Let rise until light, about 10 minutes. Bake for 10 to 15 minutes, or until browned.

POPPY SEED FILLING

"Almost any fruit can be used to fill kolačes," says Mary Victorine, "but our family prefers poppy seed, prunes or apricots. It is important to get the mixture thick enough to stay in the nest you make in the dough with your fingers."

To fill the 6 dozen kolačes in the preceding recipe, double this recipe.

MAKES ENOUGH FILLING FOR 3 DOZEN KOLAČES

½ pound ground poppy seeds
I cup water
I cup milk
I tablespoon butter
½ teaspoon vanilla extract

½ teaspoon ground cinnamon
I cup granulated sugar
½ cup crushed graham crackers
½ cup raisins, softened in water and
    drained

1. In a heavy saucepan, cook the poppy seeds and water until the mixture thickens. Add the milk and cook over low heat for 10 minutes, stirring often.
2. Add the butter, vanilla and cinnamon, stir and add the sugar. Continue cooking for 5 minutes, stirring. Remove from the heat and stir in the graham crackers and raisins. Let cool.

THE HORSERADISH
AND SPORTSMEN'S FESTIVAL
TULELAKE, CALIFORNIA, MID-JUNE

There's lots you can do with a hot root at this celebration. Kids do a three-legged Root Relay Race, hopping and handing off a hunk of horseradish. Those under five compete for Ms. and Mr. Hotsie Totsie; winners are those with costumes best representing the spirit of horseradish.

You can compete in the Tug of Root, and eat a hot dog garnished with two tablespoons of the condiment. Try the Grinding Contest; secrete yourself in an old-style phone booth with a grater and some roots, and see just how much your tear ducts can stand. The Root Toss is for kids; Root Golf is for adults. The prime rib dinner, with horseradish dressing, is for everyone.

CARROTS WITH HORSERADISH

This piquant, award-winning recipe comes courtesy of the Tulelake Horseradish Festival. When cooked, the horseradish turns nutty sweet, and its flavor complements that of the carrots.

SERVES 4 TO 6

1 pound carrots, peeled and cut into
   ½-inch slices

3 tablespoons unsalted butter

2 tablespoons freshly grated or prepared
   horseradish, or to taste

1 teaspoon fresh lemon juice
   Salt and freshly ground white pepper

2 tablespoons slivered almonds or toasted
   sesame seeds (optional)

1. Steam the carrots for about 5 minutes or until they are crisp-tender. Remove from heat and drain.

2. Melt the butter in a saucepan. Add the horseradish and stir for about 1 minute. Add the carrots and cook them for about 3 minutes. Sprinkle with the lemon juice, and season lightly with salt and pepper. Stir for another minute or so, and taste for seasoning. Garnish with almonds or sesame seeds, if desired, and serve hot.

CLACKAMAS COUNTY FAIR

Canby, Oregon

FAIR DATE: THIRD WEEK OF AUGUST, FOR 6 DAYS

GAZETTEER: *Canby is 20 miles south of Portland. The fairgrounds are on 4th Avenue.*

"Canby is the end of the Oregon Trail," says manager Barbara Lawrence, "and we are mindful of history. In 1993, which was the 150th anniversary of the Trail, wagon trains came here from Independence, Missouri, where the Trail began.

"We hold a very traditional fair. For the children we have old-fashioned games, like the three-legged race, and the egg and spoon and potato sack races. The food booths are run by the Girl Scouts and the Kiwanis and other service organizations. Nothing slick or streamlined. Admission to the fair includes the rodeo.

"The setting for the fair is beautiful; grassy grounds and tall trees. The Pioneer Village is set among the trees, and the volunteers, representing settlers and Indians, are in costume. We have five restored covered wagons there. The people in the cabins and tepees show how life was. They cook food like venison roasts and skillet bread, and some relate Indian mythology to the visitors. We gear our fair to families, and keep the atmosphere low key. People get a real sense of living history."

"On the 'Homestead Porch,' which is a false-front building, we have artifacts related to the past, and we talk about them," says Mabel Johnson. "We show a tub of wheat, to talk about how folks managed before refrigeration. Did you know that they buried their eggs, bacon and ham in bins of wheat to keep them cool? It's a fact that grains of wheat are naturally cool, and a visitor from Nebraska told me that he still takes his bedding out to the barn on stifling nights and sleeps in the wheat bin. Oats work, too, as a cooler. A very old storekeeper said that he remembered farmers coming in to sell their eggs, carrying them in buckets of oats. We show brown eggs, in a jar of water, to begin a talk about how people preserved eggs over the winter when the hens weren't laying. They'd put them in a crock and cover them with waterglass, a solution of sodium silicate that can preserve eggs for up to six months. As they age, the round yolks flatten, and

Mountain man in the Pioneer Village.

eggs lose their fresh taste. Even so, a young woman who keeps hens said she'd try it, and another visitor from Alaska, where the cost of food is so high, said she covered her eggs in melted grease to hold them. We're still using our heads today.

"People who say 'I'd love to live off the land' are nuts! In the old days they had no time for themselves; they spent most of their life preserving and chopping wood just to cook and keep warm. We are hung up on what foods are good for us—back then, food itself was what counted, and they were grateful for anything. I tell you, living off the land was tough."

MABEL JOHNSON'S SWEET AND SOUR KIDNEY BEANS

"We lived in logging camps in northern Oregon when I was a girl," says Mabel Johnson. "We didn't have refrigeration, so we'd buy just enough milk from a nearby dairy for the morning, and then buy more again in the evening.
"My mother was Swedish, and Swedes eat a lot of red kidney beans. I have to say I didn't like her beans, so I made my own recipe. This dish is cranky, in that the ingredients must be added in the order the recipe states, or it doesn't taste as good. The beans freeze very well."

MAKES 4 SERVINGS

Generous pieces of bacon (about 8 slices)	1 teaspoon dry mustard
½ cup chopped onion	2 (16-ounce) cans kidney beans
1 clove garlic, minced	1 tablespoon cider vinegar
6 tablespoons brown sugar	½ cup ketchup

1. Half-cook the bacon in a large heavy pot. Pour off all but 1 tablespoon of fat. Cut the bacon into pieces and set aside.
2. Add the onion and garlic to the pot and cook over moderate heat in the bacon fat until soft and translucent; do not brown.
3. Add the brown sugar and stir until melted. Add the mustard and stir. Add the beans and their liquid, one can at a time. Add the vinegar and stir. And the ketchup; stir. Lastly, stir in the bacon pieces.
4. Cover and cook over low heat for 50 to 60 minutes, stirring occasionally.

THE CHOCOLATE CAKE JUDGE:
Take Tiny Bites

Del Bauer recalls the Tuesday night in 1986 when, after a Rotary barbecued chicken dinner (and half his wife's meal) topped off by a chocolate nut-dipped ice cream bar, he and his friend Dennis were lassoed into being judges of the men's chocolate cake contest at the Clackamas County Fair:

"We were part of a crew of four chocolate cake judges. Now, this is serious business, and when we walked into the Kitchen Korner's demonstration area and spotted the thirty-two cakes to be judged, we knew our work was cut out for us. Dennis, who considers himself a chocolate cake connoisseur, started out taking more than a nibble of each cake. As I looked down the line, I decided to be more conservative.

"By the twenty-fifth cake, I began worrying about Dennis. I noticed he was no longer taking big bites. He was drinking more water, and there was a greenish tinge to his complexion. I looked over at the department superintendent, who was watching us and probably thinking, 'My God, I hope one of them doesn't throw up!'

"We made it, though, and started comparing notes. Thank goodness, there was complete agreement on the champion cake. It was plain chocolate with plain chocolate icing, and absolutely 'melt-in-your-mouth' wonderful.

"The last time I heard, Dennis has not tasted another chocolate cake, but I have a hunch we'll both be ready to be men's chocolate cake contest judges again—if we're asked. And if I only eat one dinner, I should be in great shape by eight o'clock, judging time."

TILLAMOOK COUNTY FAIR

Tillamook, Oregon

FAIR DATE: SECOND WEEK OF AUGUST FOR 4 DAYS, WEDNESDAY THROUGH SATURDAY

GAZETTEER: *"Cream of the Oregon Coast," Tillamook is the site of the Tillamook Cheese Factory, making a celebrated cheddar and other cheeses with the milk from 196 member dairies. The factory is one of the most popular tourist attractions in the state. The fishing and timber industries are also mainstays of the local economy, and tourism is increasingly important in this spectacular setting. Tillamook is 74 miles from Portland, on Highway 101. The fairgrounds are 2 miles east of town, on 3rd Street.*

"We have the only Pig-N-Ford races in the nation," says fair organizer Jerry Underwood, "and these have to be the funniest races you will ever see. Back in 1925, J. A. Bell and Doug Pine were driving down the Old Miami River Road in a Model T Ford, and a pig darted out in front of them. They chased it for a bit, then it darted into the brush. Suddenly another pig dashed onto the road. A great idea was born!"

Nowadays the drivers don't chase a pig, they *carry* it. Drivers, usually six at a time, pick a forty-pound pig out of a bin, tuck it under an arm, run to a waiting Model T, crank it up, and drive once around a half-mile track. They change pigs and repeat the lap, then do it a third time. The races go on for three days, culminating in the World Championship. "You need to set the spark on your Model T and lock the crank," says many-times champ George Hurliman. "Racers have broken an arm or a wrist trying to push down on the starting crank. This technique is faster." Model T owners keep their vehicles in top condition, and pass them down in the family. The races are in their fourth generation of fierce competition.

"We also have amateur boxing matches," says Jerry, "a full card. We fill the auditorium. In our courtyard there's entertainment all day; in fact, it's shoulder to shoulder."

Advance the spark!

"I've been exhibiting at the fair since I was ten. Our parents urged us to enter and to do our best," says Mildred Davy. "Since 1963 I've broadcast on KTIL radio. My brothers were amazed that I was actually getting paid to talk! During the fair, we broadcast from the courtyard. I'll interview everybody: cake contest winners, boxers, politicians, jockeys from the racetrack. It's exhausting but rewarding. My program is called 'It's a Woman's World,' and in many ways it acts as a clearing house for Tillamook. Drivers will call in and say 'I'm lost!' and I tell them how to get to a certain road. They even call us for the weather report, and ball scores. The mountains cut off radio signals from Portland, so we pretty much have a monopoly. I've traveled the world, but I wouldn't live anywhere else but a small town!"

CHEDDAR CHEESE PUFFS WITH A SURPRISE

Mildred Davy's late husband, John, worked for many years at the Tillamook County creamery. Their excellent cheddar appeared frequently on the Davy family table. If you can't get Tillamook cheese, use any fine cheddar.

MAKES 48 PUFFS

2 cups grated sharp Tillamook, or other cheddar

8 tablespoons (1 stick) butter, at room temperature

1 cup sifted flour

½ teaspoon salt

½ teaspoon paprika

48 small green stuffed olives, well drained

- Preheat the oven to 400 degrees.
1. Blend the grated cheese and butter. Stir in the flour, salt and paprika. (This can all be done in a food processor equipped with the metal blade.)
2. Mold 1 teaspoon of the mixture around each olive. Chill the puffs until firm, about 30 minutes. Arrange the puffs on ungreased baking sheets.
3. Bake for 15 minutes or until browned. (You can also freeze the puffs, well wrapped, for about 10 days. Bake them, still frozen, until browned.)

POOR MAN'S SOUFFLE

There's nothing poor at all about this dish; it should be called "Cook's Easy Souffle." For a really puffy souffle, assemble the dish and refrigerate it for 12 hours before baking.

MAKES 4 GENEROUS SERVINGS

6 slices buttered toast, cut up into cubes

2 cups grated Tillamook or other
     cheddar cheese

2 eggs, slightly beaten

2 cups milk

1 teaspoon Worcestershire sauce

1 teaspoon salt

½ teaspoon dry mustard

⅛ teaspoon freshly ground pepper

- Preheat the oven to 350 degrees.
1. Butter a two-quart baking dish. Fill it with alternating layers of toast cubes and grated cheese.
2. Beat together the eggs, milk and seasonings; pour over the toast-cheese mixture. Bake 40 minutes, or until a wooden toothpick inserted in the center comes out clean. Serve at once.

HOT CHICKEN SALAD

Mildred Davy puts out bulletins of recipes for her radio audience several times a year.
Listeners can pick them up at Station KTIL.

SERVES 4

2 cups diced cooked chicken or turkey

2 cups diced celery

2 teaspoons grated onion or 1 cup
    chopped scallion (green onion)

2 tablespoons fresh lemon juice

½ teaspoon salt

½ cup chopped almonds

1 cup sliced fresh mushrooms or drained
    water chestnuts or jicama (optional)

1 cup mayonnaise

1 cup crushed potato chips

½ cup grated sharp Tillamook or other
    cheddar

• Preheat the oven to 325 degrees.

1. Combine all the ingredients except the potato chips and cheese. Toss lightly.
   Spread the mixture in a shallow casserole dish. Top evenly with the chips, then
   the cheese. Bake 40 to 45 minutes, until heated through. The cheese should be
   bubbling.

Mildred Davy, broadcasting from the fairgrounds.

YAKIMA VALLEY JUNIOR FAIR

Grandview, Washington

FAIR DATE: MID-AUGUST, FOR 4 DAYS

GAZETTEER: *Grandview is in the fertile Yakima Valley ("The Fruit Bowl of the Nation"), approximately 150 miles from Seattle, Portland and Spokane. The area is rich in orchards and vineyards, hops, vegetable crops and dairy farms. Chateau Ste. Michelle, Washington's oldest winery, is here, as are a host of other wineries, open for touring. Visitors can also buy fruits and vegetables from farmers, and often pick them themselves. Yakima County boasts more fruit trees than any other county in America. For the fairgrounds, take exit 73 off I-82.*

"The fair's held at a good time," says Ray Vining, a fair board member. "Most of the harvest is in, and school hasn't started yet. I work the horse show. My first memory of the fair was when it was held in downtown Grandview. I recall my grandmother taking us kids to the fair with the money she earned from picking grapes. A golden memory.

"When I came back from Vietnam, I found the fair dingy; not like I'd remembered. But after a few years of working with it, I thought, 'Boy, it's neat.' It wasn't the fair that had changed, it was me; the work was therapeutic. I became president, and we moved to our new twenty-five acre site. We reused and recycled everything we could. The riding arena is fenced with old railroad ties, and we also used cedar timbers from an old slaughterhouse that had been taken up from an old road. That's double recycling. We pulled up the old flagpole and moved it; more to that than I'd thought!

"I'd rather meet people over hammer and nails than over cocktails any time. The best comes out when you're working together without any financial expectation."

"Everyone thinks of Washington as rainy," says Lucy Dykstra, "but that's on the other side of the Cascades, the mountains that divide the state. In eastern Washington we have over three hundred days of sunshine a year. We're so dry we have to irrigate, and you'll see canals everywhere. Before the railroad came, early in the century, it was all sagebrush around here. We're dairy farmers. We had a family farm on the other side of the mountains,

Denise Dykstra, ventriloquist, with Joyce and Bird, her dummies.

but it rained so much that we came to this side. I'm the dairy superintendent of the Junior Fair, which is for 4-H and FFA. It's pretty much like a larger fair, except that everything is done by the kids."

"I joined 4-H in the third grade," says Denise Dykstra, Lucy's daughter. "For my projects I show Jersey dairy cows and do photography and ventriloquism. When I was nine, I watched the Miss America pageant. In the talent portion, one lady sang, another danced, and a third did ventriloquism. I said, 'I can't sing or dance, so I'll take up ventriloquism,' and I did. My parents got me a dummy, Carrot Top, and I got books from the public library. Now I have three main dummies, and I do half-hour shows. Mom writes the dialogue. People call me. I did a show at the Kiwanis Club, and someone who saw me called for a women's club thing. I did the Elks, and a person there called me for the Masons. That was all in one week. One of my dummies is Ethel, an old granny. At the end of the performance, my throat is dry, so I put the granny, with her scratchy voice, last. I was also picked by 4-H to go to Japan for a month. Twenty-two kids from Washington were picked. Every other year, kids from Japan come here."

GRANDMA DYKSTRA'S APPLE CAKE

"It's a one-bowl cake, and it's easy and quick to put together," says Denise Dykstra, who won a blue ribbon for the cake at the Yakima County Junior Fair in a class sponsored by the Washington Apple Commission.

MAKES 1 CAKE

2 eggs	I teaspoon baking soda
2 cups granulated sugar	½ teaspoon salt
½ cup vegetable oil	2 cups peeled, sliced Washington apples
2 teaspoons vanilla extract	¾ cup chopped walnuts
2 cups flour	

• Preheat the oven to 350 degrees.

1. In a large mixing bowl, beat the eggs. Add the sugar, oil and vanilla. Mix well. Add 1 cup of the flour, the baking soda and salt and beat for 1 minute.

2. Beat in the remaining flour, then fold in the apples and walnuts. Spread the batter in a greased 9 × 11 × 2-inch baking pan. Bake for 35 minutes, or until a toothpick inserted in the center of the cake comes out clean. Cool on a rack.

LUCY DYKSTRA'S APPLE PIE IN A JAR

"These make wonderful gifts," says Lucy. "Use them as you would a can of prepared apple filling, for pies and apple crisp."

MAKES 7 QUARTS FILLING

10 cups water

4½ cups granulated sugar

1 cup sieved cornstarch

2 teaspoons ground cinnamon

½ teaspoon freshly grated nutmeg

1 teaspoon salt

3 tablespoons fresh lemon juice

Peeled and sliced apples: enough
to fill 7 quart jars

1. Make the syrup. Put the water in a large heavy pot and stir in the sugar, cornstarch, cinnamon, nutmeg and salt. Bring to the boil, stirring constantly, and continue boiling until the syrup is thick and bubbling. Add the lemon juice.
2. Fill clean mason jars with sliced apples. Cover the apples with syrup. Seal and process in a boiling water bath for 20 minutes.

Blanketed sheep, ready for the show ring.

THE GRANDVIEW GRAPE STOMP
Grandview, Washington, Mid-June

Come up to wine country, wash your feet; grab two buddies, and compete! Here's how it goes: Forty teams of three each, all over twenty-one, vie to see who can pedally press out the most grape juice in three minutes. Only one barefoot stomper at a time; each goes for one minute, squishing and squashing on fifty pounds of grapes. Shorts are recommended, and teams may wear uniform T-shirts. Past squads have included Grape Expectations, (an all-pregnant team), Footloose Fermenters, Grape Scott, and the Grapeful Dead. Officials provide a hose to wash off juice, but bring your own towels. When you're cleaned up, enjoy live music, a tasting of Yakima Valley wines, and a food fair.

EVERGREEN STATE FAIR

Monroe, Washington

FAIR DATE: LATE AUGUST FOR 11 DAYS, ENDING ON LABOR DAY

GAZETTEER: *Monroe, the gateway to the Cascade Mountains, is thirty miles northeast of Seattle in rural Snohomish County. The lush valley is ideal for raising cattle and horses and such crops as raspberries and strawberries. Fishing is excellent. To reach the fairgrounds, take U.S. Highway 2 out of Everett to Monroe.*

"The fair has good food and exciting entertainment," says manager Corey Prentice, "especially the International Lumberjack Shows. The spectators love the log-chopping competition and the races. Loggers sprint up poles, and they have to ring a bell at the top; first bell wins. They toss axes at targets, and spin a log on the pond, trying to knock the other guy off. This goes back to the old days of gathering up logs on the river. In Monroe and the surrounding area, logging was the big industry, but the spotted owl business has certainly slowed the timber industry.

"A survey showed that animals are the number-one draw of our fair. We've got draft horses, a rodeo, a milking parlor, beef animals and various breeds of swine, sheep, dogs and cats. We've got four hundred and fifteen horse stalls. The second half of the fair is Western games, and the last four days see a full-fledged rodeo. We've got auto racing, featuring the NASCAR/Winston series with Super Stocks, Bombers, Minis and Figure 8s. The seventy-five car Demolition Derby is a thriller! Our number two attraction is the food; they forget their diets and just go for it. You can also buy good American Indian food provided by the Snoqualmie tribe."

"My husband, Don, and I were honored in 1991 as the Evergreen State Fair family of the year," says Ella Walker, who worked for the fair for seventeen years, the last seven as manager. "When I retired in 1987, they named me 'My Fair Lady,'" she recalls proudly. "Don was an educator, and he's retired too; we raise French Limousin cattle at our Walking R Ranch. An entire wall in the ranch house is covered with the ribbons and trophies we've won showing the animals. Limousins have very lean meat, and they are what's called 'easy calvers,' so you don't have so many complications when they give birth. During the summer, the animals are out to pasture with plenty of water and the minerals and salt we provide, so we have more time for the fair. We have five children and nine grandchildren, some of whom show their own cattle."

ELLA WALKER'S SUGAR WORK

"I've won blue ribbons with my sugar Panoramic Easter Eggs," says Ella. "I learned sugar work from a teacher I went to to learn how to decorate cakes; I wanted to do birthday cakes for my children. The work is not really hard to do, but you should have someone show you. You fill a mold, and when the sugar formula is dry, you carefully hollow it out with a spoon, just leaving a shell. I paint the eggs, and arrange little scenes inside that you look at through a peephole. I also like to make wedding bells—two bells arranged on a plate for a 50th anniversary.

"If you are tempted to try, here is the formula for a small batch—it's just sugar and egg white: three and one-third cups sugar to one medium egg white. Stir with a spoon for two minutes, then rub the mixture between the palms of your hands so that each granule is coated with egg white. You dust a plastic mold with cornstarch, and press in the mixture. When the sugar feels dry, carefully scrape out excess with a spoon. The shells should not be more than one-eighth inch thick. You bang the mold on the table, turn it over, and the sugar should come right out."

Don and Ella Walker, 1991 Fair Family.

ELLA WALKER'S WHITE FRUITCAKE

"At Thanksgiving," says Ella Walker, "we make our Christmas cake. It's almost solid fruit and nuts. A section of glazed cake wrapped in cellophane makes a really nice present for friends. I give one to the mailman, too."

MAKES 12 SMALL (3 x 6 x 2-INCH) FRUITCAKES

1 pound (4 sticks) butter, at room temperature

2 cups granulated sugar

10 eggs

4 cups unsifted flour

1 teaspoon mixed pastry spices (or substitute a mixture of ground cinnamon, nutmeg, allspice and cloves)

½ teaspoon mace

2 teaspoons baking powder

1 teaspoon salt

2 (13½-ounce) containers of glacé red cherries

2 (13½-ounce) containers of glacé green cherries

2 (13½-ounce) container of candied pineapple

1 pound white raisins

½ pound currants

½ pound shelled walnut halves

1 cup unsweetened pineapple juice

½ cup light corn syrup

Additional nuts and cherries for decorating

• Preheat the oven to 250 degrees. Line an 18 × 12 × 2-inch pan with 4 layers of brown paper. Grease the top layer and cover it with a sheet of wax paper. (The brown paper prevents burning over the long baking period.) You can also use two smaller pans.

1. Using an electric mixer, beat the butter and sugar together on low, then moderate, speed until light and fluffy. Add the eggs, one at a time, and beat after each addition. The batter should be well mixed.

2. In a large bowl, mix together the flour, spices, baking powder and salt. Work in the cherries, pineapple, raisins, currants and walnuts.

3. Add the flour mixture to the butter mixture, and blend. Place in the prepared pan and bake for 2 hours. Cool on a cake rack.

4. When the cake has cooled enough to handle, cut it into 3-inch by 6-inch sections. Make the glaze: Combine the pineapple juice and corn syrup and bring to a rolling boil. Brush the sections with glaze, and decorate with nuts and cherries. When the glaze is cooled, and not before, wrap the sections in wax paper and then in foil. Store in tins. The cake will keep for a long time.

SALMON AT THE SNOQUALMIE LONGHOUSE

For many centuries the Snoqualmie tribe fished the Washington river bearing their name for salmon, fashioned superior traps and fishhooks, built and carved canoes with fire-hardened stone tools. Spectacular, sacred Snoqualmie Falls was—and remains—the center of their spiritual life. Each village held one or more log long-houses, structures 120 by 40 feet, that could house as many as twenty families. Then came the settlers.

By 1858, hostilities, relocation and disease had reduced head chief and shaman Pat Kanim's followers from a host of fifteen hundred to a mere three hundred souls. The chief refused to move the remnants of his people to a designated reservation, and it was that decision that would cause the tribe more sorrow. They became a tribe without land, and hence, in the eyes of the Bureau of Indian Affairs, no tribe at all.

That has changed. After lengthy battles, the Snoqualmie have finally proved that they are indeed a tribe, and are recognized as such by the Federal Government, state of Washington, and other Indian tribes. Though landless, the five-hundred strong Snoqualmie survivors have established legally what they have always known in their hearts: that they are truly a tribe. "We've had to *show* that we are a tribe, that we have continuity," says Kathy Barker, secretary of the tribal council. "It's never too late to be re-recognized," adds Kathy's daughter Shelley Burch, who is vice-chair of the Snoqualmie tribe.

"We started smoke-barbecuing and selling salmon at the Evergreen State Fair in 1983, as a fundraiser," says Shelley. "We worked out of a small booth. Then next year, through the hard work and dream of Mr. Walt Crane and many others, we built a cedar longhouse. A hundred can fit into it easily. As well as food, we sell our arts and crafts, and use the money for education and to help the needy of the tribe. This area is called Frontier Land, and totem poles frame the entrance and line one side of the longhouse. We've built a barbecue pit, and we cook salmon over alder wood. The wood must not be too green or too dry if it is to burn properly. Alder gives a very special flavor that's hard to describe: sort of tangy but mellow, I'd say. Not at all like mesquite or hickory.

"There are totem pole carving contests, and members of woodcarving groups compete."

Snoqualmie tribe salmon stand.

SHELLEY BURCH'S SMOKE-BARBECUED SALMON

"You will need about 20 pieces of split alder wood," says Shelley. "Serve the salmon with coleslaw and baked beans or corn on the cob. "These are the salmon that may be used: king, coho, sockeye, humpy, and pink."

SERVES 12-24

1 pound (4 sticks) butter	Pepper to taste
1 tablespoon garlic salt	5 to 10 pounds salmon fillets, from
¼ cup parsley flakes	salmon weighing 3 to 5 pounds

1. Build a small fire of alder wood in a barbecue pit, adding pieces to the fire as needed to maintain a blaze to build up a bed of hot coals.
2. Melt the butter with the garlic salt, parsley flakes and pepper. Brush the mixture over the fish. Place on a grill over the coals. Cook, turning once. You will know when the salmon is done when the white, pearly juices rise. Over hot coals, cooking time should be about 10 minutes per side.

SHELLEY'S SALMON SOUP

"Serve with a roll or any type of bread," says Shelley Burch.

SERVES 8

3½ quarts water	1 small bay leaf
1 to 1½ pounds boned salmon pieces, about 4 ounces each	5 strips bacon
	¼ cup chopped onion
1 teaspoon salt	4 tablespoons (½ stick) butter
3 medium potatoes, peeled and cut into small pieces	½ cup chopped celery

1. In a large kettle, combine the water, salmon pieces, salt, potatoes and bay leaf. Bring to the boil, then lower the heat to a simmer.
2. In a skillet, brown the bacon. Remove it and drain off most of the grease. Add the onion, butter, and celery and cook until soft and translucent. Add the bacon, crumbled, and the onion and celery to the soup pot. Cook the soup slowly, partially covered, for 35 to 40 minutes. Stir occasionally. Remove the bay leaf before serving.

ALASKA STATE FAIR

Palmer, Alaska

FAIR DATE: LATE AUGUST FOR 11 ^{1}/2 DAYS, ENDING ON LABOR DAY

GAZETTEER: *Palmer, calling itself "Alaska at Its Best," is a small town 42 miles northeast of Anchorage, and the trade center for the Matanuska Valley, a fertile growing area and seat of tourism and recreation. Visitors golf, canoe, ride horseback, observe wildlife such as musk oxen, and revel in breathtaking views of the Knik Glacier and snow-capped mountains. The fairgrounds are located at Mile 40 Glenn Highway, 6 miles from the Parks Highway intersection.*

"We are the prime agricultural area of the state," says fair manager Sara Jansen, "and we've never lost sight of our agricultural beginnings. The fair began n 1936, as a celebration and a respite from work by the colonists. These were two hundred and two families from Michigan, Minnesota and Wisconsin that the federal government brought to Alaska in 1935 to make the raw land bloom, and also to give these homesteaders a chance to succeed in the midst of the Great Depression.

"Our fairgrounds are in the most beautiful natural setting you can imagine. Driving to work, I look around at the mountains and think 'another day in Paradise.' We've moved some of the original colony buildings onto the grounds, including one of the three original churches, that's used year-round as a community theater. We have over seven hundred events, including a potato digging contest, an ax throw, and a greased pole climb for kids under twelve; there's a fifty-dollar bill at the top, so of course this is a popular event! We include our native peoples, too; we have

Winning vegetables.

beading, skin sewing and basketry events, and we've had native dance troupes, including Siberian Eskimos.

"The Diaper Derby is an event that always cracks me up. It's for our Alaskans under eighteen months. A mom or dad is at the starting line, with another family member waving a toy at the finish line. Generally, no one moves. A baby may break out of the pack and begin the fifty-foot crawl, but it never fails that the child sits down two feet from the finish line.

"Our fair is famous for its giant vegetables, mainly cabbage. With great top-soil and twenty hours of summer sun, you can grow big vegetables! I believe the record cabbage is just over ninety-eight pounds. Most of these cabbages are used for sauerkraut. Surveys say that people come to the fair from all over the state to be with friends. Alaska, we say, is really a small town."

"The fair is the social event of our short summer, " says Sue Anne Weaver, who is supervisor of special contests. "My husband is a biology teacher, and I tutor migrant kids. Unlike the Lower Forty-eight, where migrants are usually farm workers, ours are fishing families who move to various sites and camps. I counsel and advise these kids. We lived in the bush, in a small village on the Yukon River, for three years when my husband was teaching there, and the food had to be flown in by light plane. A chicken could cost five dollars, but a whole local salmon only cost one dollar. We had salmon four days a week for those three years."

SUE ANNE WEAVER'S SALMON

"I've had my fill of salmon," says Sue Anne, "but I know how to cook it. This recipe is simple but good; an Eskimo village elder gave it to me. The mayonnaise keeps the fish moist."

SERVES 4 OR MORE

1 or 2 boneless salmon fillets	Mayonnaise
Salt	Lemon pepper seasoning

* Preheat the oven to 350 degrees.
1. Cover a baking sheet with lightly oiled foil. Place the salmon, skin side down, on the foil. Salt it lightly and cover evenly with mayonnaise. Sprinkle on the lemon pepper seasoning.
2. Bake in the center of the oven for 30 to 40 minutes, until the salmon is opaque all the way through and is beginning to flake at the thickest point.

SUE'S ALASKAN-CAJUN BLACKENING MIX

"We used this seasoning mix mostly for halibut and salmon, but I've also used it to coat fried clams and steak. I got the recipe from my neighbor. Sometimes you have to talk people into trying the first bite, because the stuff looks burnt! However, I've yet to meet the person who hasn't loved it once they've tried it."

MAKES ABOUT 1/2 CUP

1 tablespoon paprika	1 teaspoon ground white pepper
2½ teaspoons salt	1 teaspoon ground black pepper
1½ teaspoons onion powder	1 teaspoon crumbled dried thyme leaves
1½ teaspoons garlic powder	Halibut or salmon fillets
1 teaspoon ground red pepper	Melted butter

1. Mix the first 8 ingredients thoroughly.
2. Cut fish into 1/2-inch-thick fillets. Pour melted butter into one shallow bowl, the seasoning mixture into another. Turn the fillets in the butter, then dredge them in the seasoning mixture. (Messy, but really coats everything real well.)
3. Heat a cast iron skillet until almost red hot, or heat it outside on a grill. (We do this even in winter, since cooking blackened fish is a smoky process.) Cook the fillets, not crowding the pan, until black on one side; turn, blacken the other side.

MAUI COUNTY FAIR

Kahului, Maui, Hawaii

FAIR DATE: OCTOBER, THE WEEKEND BEFORE COLUMBUS DAY, FOR 4 DAYS

GAZETTEER: *The fair is held at the corner of Kanaloa and Kaahumanu avenues, in Wailuku, the county seat.*

"Everybody on Maui goes to the fair," says publicist Jan Trojan. "It has a long tradition, and it's exciting for the island. In the old days, the fair president was as important as the mayor. You can see the fairgrounds as you get off the boat, or taxi drivers will shuttle you from the airport."

"We had a disastrous Kona storm from the south about twenty years ago," says adviser Bob Jones, "so we changed the date to when the weather's better. Before 1965, they'd let the school kids out, and they'd come to the fair on sugar cane trains from various plantation camps. When the fair started over seventy years ago, sugar cane was the only crop, the only industry. The workers came in waves—Chinese, Japanese, Portuguese and more recently from the Philippines. New groups put up camps, like army bases. They're abandoned now, but you can tell where they were; you'll know by the groves of avocado, mango and monkey pod trees out in the middle of nowhere.

"Agriculture has diminished, and tourism has grown. Our fair's switched gears from agriculture, and has become ethnic, in that all the groups on the island can show their food, their lifestyles, and raise money. We have thirty-two food booths, and all the food is different. For years the Catholics have had hamburgers, and the United Church of Christ handles soda. Each new group must find a new food. We have authentic Chinese chow fun noodles,

Ikue Arisumi, with giant daikon radish.

Korean ribs, Philippine adobo chicken, and codfish cakes from the Puerto Rican Club. The Maui Swim Club serves Loco Moco, a hamburger patty with a fried egg and gravy, usually served with rice, for breakfast.

"There's a special day at the fair for people with mental and physical disabilities who might feel intimidated visiting the fair during regular hours. A special time has been set aside for them. Of course, they're welcome any time, but with the volunteers and care givers, it's easier for them.

"Our fair is more for the backyard gardener, because sugar and pineapples aren't involved. At Orchidland, growers and clubs set up areas like small gardens, about ten feet square, with moss, ferns and running water. We have competitions for corsages and leis. The tropics are ideal for orchids; just create a little shade and you have a greenhouse.

"With only four high schools on the island, you can imagine how short the football season is. Six games! The kids don't get much chance to crank up their cheering, so we put them on a football field and they can scream their heads off.

"The fair's a great place to see the different races, and to try to figure out what people are. The Hawaiian-Chinese-Portuguese girls are just beautiful. We have so many visitors from everywhere, too. We have a map of the world, and stick in pins for visitors. We've had them from forty states, Japan, Germany, and France."

"I co-chair the Homemaker's Display," says Phyllis Takeuchi, "and I've been fifteen years with the fair. I was born on Oahu, and I'm the third generation in Hawaii. My grandparents were plantation workers, and my father was an electrician. Education was so important to them; my brothers and I finished college. I teach elementary school. Forty years ago, when I went to school, it was mostly with Japanese; now it's much more of a mixture of East and West.

"At the fair you learn about different foods, and ask for recipes. Office groups, churches and family

gatherings try them out. People like to make connections, form a closeness. We get together, visit, and we call that 'talk story.'"

KIM CHEE POKE IN A RICE PAPER TACO WITH AVOCADO SAUCE

This winning recipe from the Maui County Fair is a small, contained lesson in Hawaiian cooking, which is a lively fusion of culinary cultural influences. Some of the ingredients need translation or clarification: Poke—pronounced po-KAY—is a raw seafood appetizer, made in many ways. Ahi is yellowfin tuna. Maui onions are celebrated for their mildness. Rice paper is widely used in Asian cooking. Won bok is mostly called nappa cabbage or Chinese cabbage on the mainland, and Hawaiian salt is coarse, like kosher salt.

Kim chee is the piquant Korean pickle made (mostly) of Chinese cabbage and hot peppers, appearing often at Korean meals. But on Maui, kim chee is more than a condiment; it's almost a way of life, and islanders of all ethnic backgrounds create new recipes for it all the time. In fact, they love their kim chee so much that the fair sponsors a kim chee contest, with two divisions: traditional and Maui style. Joint 1993 winners of the Maui-style version were John Peck and Vu-Tran, of the Villa Restaurant in the Westin Maui hotel.

SERVES 6

FOR THE KIM CHEE:

¼ cup Hawaiian salt	1 cup finely julienned carrots
8 cups water	2 teaspoons raw sugar
2 pounds won bok (nappa) cabbage, roughly cut	⅓ cup roughly cut scallions (green onions)
4½ teaspoons chili paste	¼ cup roughly cut fresh chives
4½ teaspoons minced fresh garlic	2 tablespoons shrimp sauce
⅓ cup oyster sauce	2 teaspoons grated fresh ginger

FOR THE POKE:

2 cups kim chee (above)	⅓ cup shoyu (soy sauce))
12 ounces raw ahi (yellowfin tuna), chopped large	1 tablespoon oriental sesame oil
	Oil for deep frying
1 medium Maui onion, diced	6 sheets rice paper

FOR THE AVOCADO SAUCE:

2 ripe dark-skinned Haas avocados, peeled and diced	2 tablespoons lemon juice
	1 tablespoon chopped cilantro
¾ cup chicken stock	Salt and white pepper to taste
½ cup sour cream	

1. Make the kim chee: Add the salt to the water, add the won bok and soak for 2 hours. Drain the mixture and add the remaining kim chee ingredients. Let stand for 2 hours.
2. To make the poke: Chop 2 cups of kim chee into fine pieces and mix with the ahi, onion, shoyu and sesame oil.
3. Make the "taco" shells. Pour 4 inches of oil into a deep skillet or pot and heat until bubbling. Toss in a rice paper sheet. Use a large metal whisk to push the sheet down in the oil. The rice paper will form itself into a taco shell around the whisk. Fry 45 to 60 seconds; the rice paper should be translucent and bubbled. Remove and drain. Proceed with the rest of the rice paper.
4. Make the avocado sauce: Combine all the ingredients and whirl in a blender or food processor until smooth. (To keep the sauce from darkening, place a sheet of plastic wrap directly on the sauce until needed.)
5. Fill the shells with the kim chee mixture and serve with avocado sauce.

PHYLLIS TAKEUCHI'S HAUPIA (HAWAIIAN COCONUT PUDDING)

If you can find only unsweetened canned coconut milk, up the sugar to 1 cup.

MAKES 1 PAN PUDDING

1 quart sweetened coconut milk (can be canned)	¼ cup granulated sugar
¼ cup sieved cornstarch	Pinch salt

• Place all the ingredients in a heavy saucepan and cook, stirring, over moderate heat until the mixture thickens. Pour into an 8-inch-square glass dish; refrigerate when cool. To serve, cut into squares.

THE 442ND VETERANS CHICKEN BARBECUE

Tangy grilled chicken is the specialty sold at the booth run by men who served with great distinction in World War II. Their group, the all Japanese-American 442nd Regimental Combat Team, served in France and most particularly in Italy.

"We were *nisei*, second generation, from Hawaii and the Mainland," says Toshio Endo. "We formed the 5th Battalion of the regiment. We trained at Camp Shelby in Hattiesburg, Mississippi, in '43 and '44, and were shipped to Italy, where we joined the 100th Battalion, which was there already." The 442nd was among the most widely decorated units in the U.S. Army: three thousand six hundred Purple Hearts for wounds received in action, three hundred and fifty-four Silver Stars, forty-seven Distinguished Service Crosses, one Distinguished Service Medal and one Medal of Honor. Some eight thousand Japanese-Americans served.

"Years ago we started at the fair selling goldfish," says Toshio. "We issued people little paper nets, and they could fish in our tank until the nets broke. The price was fifty cents. Chicken makes more money!

"We sell about seven thousand three hundred pounds of chicken thighs, clamped into screened grills and cooked over kiawe wood, a wild hardwood tree that grows on Maui and makes excellent fires. The chicken is marinated in a sauce of brown sugar, shoyu sauce, garlic and ginger. We huli huli the grills; in Hawaiian, huli means to turn. We turn the screens all the time so the good juices don't drip away. The barbecue aroma goes around the fairground and really builds up the appetite!"

442nd Regimental Combat Team veterans.

442ND VETERANS CHICKEN

Here is an edited-down version, for home use.

SERVES 8 TO 10

1 cup packed brown sugar
¼ cup pulverized fresh garlic
¼ cup pulverized fresh ginger
2 cups shoyu (soy sauce)
4 pounds chicken thighs, trimmed

1. Combine the sugar, garlic, ginger and shoyu in a large bowl. Marinate the thighs for 1 to 2 hours. Pat dry.
2. Cook over white-hot coals for about 30 minutes or until tender, turning often.

NINE

THE AGRICULTURAL ORGANIZATIONS

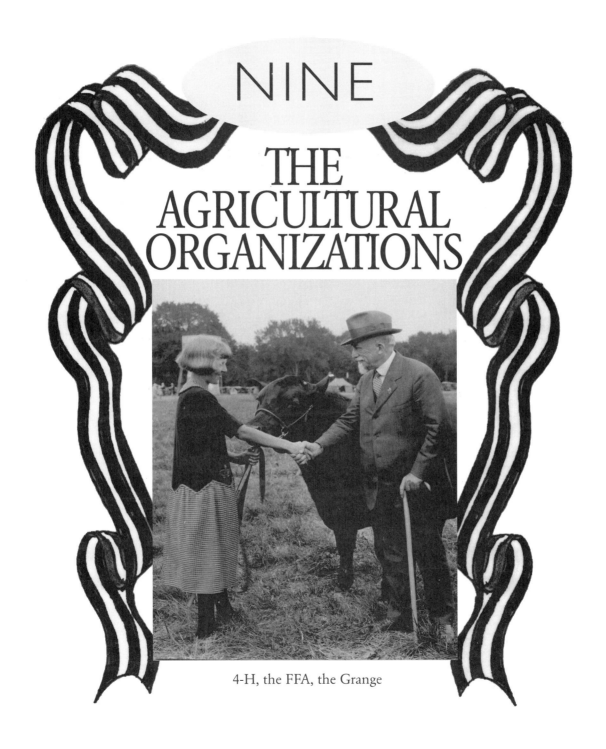

4-H, the FFA, the Grange

THE 4-H

At the turn of the century, the Department of Agriculture and worried Midwestern educators feared that city glitter was luring too many young people away from the soil. Innovative farming programs aimed at youth were tried. A corn-growing contest, with a one-dollar prize offered by a seed developer, was a thumping success. Canning, poultry and other agricultural clubs caught on.

In the early 1900s, what is now known as 4-H began in a number of places at the same time. In Ohio, for example, school superintendent A. B. Graham organized agricultural and home economics clubs for boys and girls. Similar clubs sprang up in other states. These clubs were the forerunners of the 4-H movement. (The clubs had an important impact on agriculture: the USDA felt the best way to persuade farmers to adapt innovative farming methods was to go through their children.)

In the early 1920s the clubs amalgamated, adopting the green four-leaf clover symbol, with the letter "H" in white or gold on each leaf. In 1927 the 4-H pledge was approved: "I pledge my Head to clear thinking, my Heart to greater loyalty, my Hands to larger service, and my Health to better living—for my club, my community, my country." In 1973, the words "and my world" were added to the pledge.

4-H is administered by the Cooperative Ex-tension Service, which is the educational arm of the United States Department of Agriculture, operating out of each state's land grant universities. The Morrill Act of 1862 provided federal land for institutions offering programs in agriculture, home economics and engineering in every state. "In 1890," says Dr. Alma C. Hobbs, assistant deputy administrator for the Extension Service's 4-H and Youth Development, "institutions for blacks were established in segregated states. Therefore Southern states have two land grant universities. The Smith-Lever Act of 1914 established the Extension Service at the land grant colleges and universities. Each campus has an office for Extension Service, and they provide leadership for each county Extension office. 4-H is in every county in every state. It is the largest youth organization in the country, with over five and one half million members."

As the country's demographics have changed, so has 4-H; only 12 percent of its members now live on farms. Everyone is welcome, though, and 24 percent of the members are from minority groups. Each member must choose at least one project per year, subjects range from raising corn for a farm kid, to raising cherry tomatoes in window boxes for an inner-city youngster, to community service. The 4-H aim is to encourage young people to become competent, coping and caring in all that they do. "We originally had members aged ten to twenty-one," says Dr. Hobbs, "but now we are getting them as toddlers. We are making 4-H inclusive, not exclusive. We are accessible to all youth, not just middle-class white children."

Not surprisingly, several secretaries of agriculture have been 4-H'ers. Alumni include Vice President Albert Gore, opera singer Sherrill Milnes, Heisman Trophy winner Herschel Walker, singer Reba McEntire, entertainer Roy Rogers, astronauts Alan B. Shepard, Jr., and Ellison S. Onizuka, and former First Lady Rosalynn Carter.

THE 4-H GIRLS: RAISING GRAND CHAMPIONS

"I'm really happy about Kim and Stacy's involvement," says Gay Quarty, mother of two now-grown 4-H'ers. "We moved to Newton, New Jersey, to get away from the fast life. When we came here years ago, there were more cows than people. The 4-H is very big in this area; some kids go into sewing, some into dairy animals or rabbits—take your choice. My girls decided they'd like to raise beef cattle. The county representative recommended Polled Herefords, brown and white cattle without horns. He said they were more docile than Angus and other breeds—at full growth, those animals weigh about a thousand pounds, so you can believe gentleness was a consideration for my ten- and eleven-year-old girls.

"4-H is a big commitment. Children have to take care of their animals every day; no skipping feedings or grooming, being sure to get a steer accustomed to a halter and to walk and stand for the judges. A kid who tries to get the family to help, it shows at the county fair when a child leads an animal that doesn't trust or know him into the ring, and the steer acts up.

"All the children who compete at fairs have animals of the same age: they'll get a calf about five months old and keep it for a year and a half. Once a month the girls went to a meeting at a farm, with demonstrations on training, clipping, grooming, how much hay and grain to feed for proper weight gain, even how to wash the animals with soap and water, and how to polish their hooves. The animals developed a bond with Kim and Stacy—we really had a lot of fun."

Kim Quarty reflects on her steer raising and competitive showing days as a diligent 4-H'er. "At first Stacy and I bought four- or five-month-old-calves in the spring. We liked picking them out, but even in selecting the little animals there was competition. With years of experience my sister and I learned to pick out a really good one, and that made the competition for us. Later, we bred our own calves. But the spirit of 4-H is really to help each other, more than outright competition. We'd start in the spring going to county fairs, and as you won, you'd move up a level.

"At about the age of one and a half, the steer is considered 'finished'; that is, ready to be bought for meat. Raising a steer is what 4-H called a market project, so a sale is necessary. My last year of showing I had two fine steers, Howard and Trevor. Howard won at the Sussex County Fair in August: he was Grand Champion, so I had to sell him: that's a rule. I took Trevor to the New Jersey State Fair at Flemington that October, and we won Grand Champion.

"It's hard, at first, for a kid to face the fact that the animal she's raised will be slaughtered for meat, but that is, really, after all, the purpose. And that animal has had the best care and life that was possible.

"When we first started, we raised a steer named Oliver, sold half of him to a

Stacy Quarty.

Kim Quarty.

farmer, and kept a portion of the meat for ourselves. I just couldn't eat Oliver. But after a few years, you get used to it. As my mother said, there would always be another animal to care for. You learn to be practical."

BEEF STEW A LA OLIVER, WITH OLIVES

Gay Quarty named this stew in honor of Oliver, one of the first steers her daughters Kim and Stacy raised.

SERVES 6

8 slices medium-cut bacon	Freshly ground black pepper
2 pounds beef chuck, cut into 2-inch cubes	1 bay leaf
1 teaspoon salt	6 medium carrots, peeled and cut into
½ cup dry red wine	2-inch lengths
½ cup water	2 medium onions, peeled and quartered
3 cloves garlic, chopped	1 (10-ounce) package mushrooms,
¾ teaspoon dried thyme leaves, crumbled	washed, trimmed and sliced
¾ teaspoon dried rosemary leaves, crumbled	1 cup pitted black olives, rinsed

• Preheat the oven to 350 degrees.
1. In a heavy casserole large enough to hold the stew, fry the bacon until crisp. Set aside to drain on paper towels.
2. Dry the cubed meat well and sprinkle with the salt. Brown it in the hot bacon fat in batches, turning often. Set aside.
3. Drain off all the fat. Over moderately high heat add the red wine, and as it bubbles up, scrape the browned bits from the bottom of the casserole; they add great flavor to the sauce. Return the meat to the casserole and add the water, garlic, thyme, rosemary, pepper, and bay leaf. Bring to the boil, cover, and transfer to the oven. Cook for 1 hour, checking occasionally to stir and to see that the pot simmers, not boils. If it does, reduce the heat. (You may also cook the stew on top of the stove over the lowest possible heat.)
4. Add the crumbled bacon, the carrots, onions, mushrooms and olives. Bring to the boil on top of the stove, lower the heat and simmer until the beef and vegetables are tender, about 45 minutes. Remove the bay leaf and serve in deep bowls with warmed French bread for dipping.

"YE OLDE BARNYARD SHOW"

Dan Hurld, Jr.,'s mission is to champion the farmer to people who encounter his product only as neatly wrapped packages in the supermarket. "When I was a young Extension Agent just out of college in 1952," says Dan, "I could see that fair visitors didn't know a thing about the business of agriculture, and how the farmer keeps us fed for practically nothing. I went to a fair manager and said, 'You need to tell city folks about animals.'"

In Ye Olde Barnyard Show, Dan manages to present the nuts and bolts of agricultural economics in an entertaining way. He has enlightened and delighted audiences at fairs throughout the northeastern states, and from Vancouver to Halifax in Canada. Dan doesn't pull punches. He makes it plain that our chicken dinners start with live animals, and that someone's got to dispatch, clean and package the critters.

"'The Infamous Chicken Chop' is one of eight shows I do. I have a live chicken in a little cage, and I play a lot of different parts. To start, I come out and heft a big cleaver, shake it toward the chicken, and I ask for helpers from the audience. I need a gizzard grabber, a feather puller, a chicken plucker. and someone to hold the chicken. What I'm trying to illustrate is that a lot of people work to get you that chicken all nicely ready for the oven. If you bought a live chicken on the hoof, you wouldn't pay much—but do you want do all those messy jobs? I give marketing lessons. I'm the chicken raiser, bargaining with the grain man who wants a lot of money; I'm the buyer, who tries to beat down the farmer; I'm the man delivering the chickens to the plant, where the workers are on strike, but the SPCA says I have to get those chickens in pronto; and then I'm the one who has to cut up those chickens. Meanwhile, I'm putting my helpers through a practice run. The audience starts to yell and groan, 'Don't kill that chicken!'

"The point of my little show is that you, the consumer, pay more for electricity, cleaning up, transportation and packaging than you do for the actual animal the farmer has raised. So do me a favor and don't complain about the price. Thank the Lord that we have productive farmers raising food of high quality for very little. And if you want others to do the preparation work for you, you're paying them, not the farmer. If you agree, I tell the audience, I'll let this chicken go! And, of course, they do!"

Dan and Barbara Hurld, 1970. Dan drives Arabian mare Rubiyat, hitched to a Meadowbrook cart.

CHICKEN ROSEMARY

"Easy as can be, but good enough for company," says Barbara Hurld.

SERVES 4

1 (3-pound) chicken, cut up, or equivalent parts	½ teaspoon crumbled thyme leaves
Salt to taste	1 chicken bouillon cube, dissolved in ⅓ cup boiling water
¾ cup all-purpose flour	1 teaspoon lemon juice
1 teaspoon paprika	1 tablespoon dried rosemary leaves

- Preheat the oven to 375 degrees.
1. Rinse the chicken pieces and pat dry. Salt each chicken piece. In a bag, combine the flour, paprika and thyme, and give the bag a shake to mix the ingredients. Put the chicken pieces in the bag, a few at a time, and shake to coat.
2. Arrange the chicken pieces, skin side down, in a lightly oiled baking pan. Bake for 30 minutes, or until tender.
3. Turn the pieces over. Baste with the bouillon broth and the lemon juice, and sprinkle on the rosemary leaves. Cook an additional 30 minutes, basting twice.

BARBARA HURLD'S WITCH'S BREW SALAD DRESSING

"We had some friends over," says Dan Hurld, "and Barbara was over by the sink, making her salad dressing. Our friends commented that she looked like she was whipping up a potion or witch's brew, hence the name."

MAKES ABOUT 2 CUPS

⅓ cup granulated sugar	1 teaspoon paprika
1 teaspoon dry mustard	¼ cup finely chopped onion
1 teaspoon celery seed	¼ cup cider vinegar
1 teaspoon salt	⅓ cup ketchup
½ teaspoon garlic powder	1 cup vegetable oil

1. Place all the ingredients in a quart jar. Cap it and shake well, or emulsify the dressing in a blender. Store in the refrigerator for up to 1 week.

THE NATIONAL FFA ORGANIZATION

"The goal of the FFA," says Dr. Larry Case, CEO and national adviser to the organization, "is to prepare young people for careers in the science, business and technology of agriculture. Members of the FFA are students enrolled in vocational agriculture classes in high school, and each school that offers these courses has an FFA chapter. In other words, the schools don't 'teach FFA,' but the formal education makes it possible for students to join FFA in their freshman year, and to remain members until the age of twenty-one, benefiting from the guidance and help of teachers and counselors."

FFA (Future Farmers of America) was organized in Kansas City in 1928, as an outcome of the Smith-Hughes Act of 1917, providing federal funds, to be matched fifty-fifty with participating states, for the promotion of vocational education for boys in public high schools. Each chapter has its annual activities program, with all members participating. Such skills as public speaking and understanding parliamentary procedure are encouraged, along with practicalities like swine management or improved hay production. A chapter might own a quality hog sire so members can upgrade the quality of their stock; in an urban area, a chapter might own a greenhouse for flower growing. Although 4-H and FFA are separate organizations, many young people belong to both.

This is an organization with uniforms, rituals and titles. Members begin as Greenhands, and through achievement rise to Chapter Farmer, State Farmer and American Farmer. FFA understands full well that the future will not hold too many small farms, but agriculture—agribusiness—is an economic giant, employing 17 percent of the American work force. "Ag ed," as agricultural education is known, covers the bases, dealing with wildlife management, animal science, horticulture, marketing and all areas of agribusiness. Membership is around half a million. Girls have been admitted as full FFA members since 1969.

"We've got six high schools here in Clark County, Ohio, and five of them have

ag ed courses," says Betty Kitchen, who works with the Clark County Soil & Water Conservation District, which sponsors a soil judging contest in the county. She has also judged. "Actually," says Betty, "soil judging takes place across the country. Among all the FFA competitions, I'd say soil judging, for us, is the biggest. As prestige, it's right up there with baseball and football. About a hundred kids participate. We pick a site, and someone with a backhoe digs three pits about five feet deep. Where the pits are to be is kept a big secret; we don't want anyone sneaking a look ahead of time! Teams of three go down in the trenches and see the various layers of soil. They've studied soil, and they should be able to tell if it's good for agriculture or not, if it will provide good drainage, or be suitable for urban development. The top individual gets a trophy, and team scores are combined. The top team goes to the district competition, and the winner goes to the state, and eventually to the national judging in the Midwest. The same judging rules apply to poultry, calves and so forth. "When an all-girl team wins—and they do!—the boys don't like that one bit.

"Certainly all FFA kids won't end up on the farm," says Betty, "but what they've learned will make them better citizens, smarter consumers, able to pick out the best property, and give them the knowledge of where food comes from."

BETTY KITCHEN'S RAISIN PIE

Betty's involvement with FFA springs from her work with the Clark County Soil & Water Conservation District in Ohio, which, in addition to sponsoring soil judging, also sponsors the hay show at the Clark County Fair. "There are different classes," she says, "such as hay baled wet or green. A man from Ohio State University judges the hay, and he tastes each sample of silage—that's the hay in its fermented form."

MAKES 1 PIE

2 cups raisins

2 cups water

½ cup packed brown sugar

2 tablespoons cornstarch

1 teaspoon ground cinnamon

⅛ teaspoon salt

1 tablespoon cider vinegar

1 tablespoon butter

Pastry for a 2-crust 8- or 9-inch pie

- Preheat the oven to 400 degrees.
1. Boil the raisins in 1³/4 cups of the water for 5 minutes.
2. Mix together the sugar, cornstarch, cinnamon and salt. Add the 1/4 cup of water and pour the mixture into the pot with the raisins. Bring to the boil. Stir in the vinegar and butter.
3. Pour the mixture into the unbaked pie shell, flute or crimp the edges and cut 2 or 3 slashes near the center for steam to escape. Bake for 35 to 40 minutes, until the top is golden brown.

BROCCOLI SALAD

"I think this salad's a bit different from other recipes you might receive," says Betty Kitchen. "It's a strange combination, but when I served it to the Clark County Fair Board, they loved it. And I made a ton of it when I had my catering business." Betty uses raw broccoli, but you may prefer to steam it lightly.

SERVES 6

1 bunch broccoli, cut into bite-sized pieces	1 small onion, chopped fine
¾ cup sunflower seeds	1 cup mayonnaise
1 cup raisins	1 tablespoon sugar
½ pound bacon	

1. Steam the broccoli for several minutes or leave it raw, as you prefer. If you steam it, run it lightly under cold water to cool it and set the color. Place in a serving bowl.
2. Add the remaining ingredients and mix well. Chill well. Spoon onto 6 plates and serve as an appetizer salad.

THE NATIONAL GRANGE OF THE ORDER OF PATRONS OF HUSBANDRY

Just after the Civil War, Oliver Hudson Kelley was asked by the Commissioner of Agriculture to travel through the South, to determine what it would take to heal the farmers' wounds. Kelley was a member of the Masons, and wore a Masonic ring. Despite resentment toward Northerners, he was welcomed into the homes of fellow Masons, who recognized the ring.

Back in Washington, Kelley reported that he felt a brotherhood, a fellowship or a club, was needed to bring the folds of North and South together. In 1867 the Grange (the name comes from a common English word for farm) was founded, a fraternal and ritualistic order. Members can rise from the First Degree of the Order in the Subordinate (local) Grange, through the Pomona (county) Grange, on through the State and up to the National Grange for the Seventh and highest Degree. The Order of Patrons of Husbandry (the Grange's fraternal name) has, from its inception, admitted women as full and equal members, the first large American organization to do so. In fact, any person over the age of fourteen can become a full member.

The Grange's purpose has been to educate and to protect rural Americans. Although it takes no partisan stand and endorses no candidates, the Grange is a strong lobbying organization. Members are urged to express opinions on community and national issues which could influence national Grange policy. Among the great victories of the late nineteenth century were the establishment of the cabinet-level Department of Agriculture; regulating the power of the railroads; and implementing rural free delivery and the parcel post system. More recent triumphs are federal crop insurance, soil conservation programs, and pesticide laws and regulations.

Grange Halls serve as community centers, and Granges organize fairs, such as the Washington County Fair in Rhode Island (page 28) and the Hookstown Fair in Pennsylvania (page 68). Grangers, as they're called, work for

Riverside Grange Hall, St. Joseph County, Michigan, formerly a church.

deaf awareness and many community causes, but they also have a considerable amount of fun; most often family fun. "There are about three thousand Grange Halls," says Judy Massabny, director of information at National Grange headquarters in Washington, "many of them a hundred years old. They've been repaired, and remodeled, and now have indoor plumbing! As suburbs and rural areas merge, I foresee that many newcomers will join. People want to become involved in community affairs. The Grange should continue to do well in the future."

RECIPES FROM THE GRANGE

All recipes are taken from *The Glory of Cooking* (1986) by the National Grange, "in which we promote the use of American agricultural products." The book is dedicated to the bicentennial celebration of the signing of the Constitution. Grange members throughout the country contributed recipes, as did the President and a number of state and federal high officials.

ONION-WINE SOUP

This recipe was given by then First Lady Nancy Reagan.

MAKES 6 TO 8 SERVINGS

5 large onions, chopped	1 tablespoon wine vinegar
4 tablespoons (½ stick) butter	1 teaspoon granulated sugar
5 cups beef broth	1 cup light cream
½ cup chopped celery leaves	1 tablespoon minced parsley
1 large potato, peeled and sliced	Salt and pepper to taste
1 cup dry white wine	

1. In a large heavy soup pot, cook the onions in the butter over moderate heat, stirring constantly, until the onions are soft and translucent. Add the beef broth, celery and potato. Bring to the boil. Lower the heat, cover, and let simmer for 20 minutes.
2. Puree the soup in batches in a food processor or blender; return to the soup pot. Stir in the wine, vinegar and sugar. Bring to the boil, lower the heat and simmer for 5 minutes.
3. Off the heat, stir in the cream, parsley and salt and pepper. Heat gently to serving temperature, but do not let boil. Ladle into soup bowls.

WILTED LETTUCE SALAD

Donna Cross of the Hawks Mountain Grange, Vermont, contributed this salad. It is a favorite in Pennsylvania Dutch country.

MAKES 6 SERVINGS

8 cups torn leaf lettuce	½ teaspoon salt
6 slices bacon	¼ cup water
½ cup chopped scallions (green onions)	6 radishes, thinly sliced
½ cup cider vinegar	1 chopped hard-boiled egg
4 teaspoons sugar	

1. Place the lettuce in a large serving bowl. In a large heavy skillet, cook the bacon until crisp. Drain the bacon on paper towels, crumble it, and set aside.
2. Pour out all but 3 tablespoons of dripping. Over moderately high heat, cook the scallions in the dripping. Add the vinegar, sugar, salt, bacon and water. Bring to the boil, stirring constantly. Immediately pour over the lettuce, toss to coat, top with the sliced radishes and egg, and serve at once.

LUCY'S CORN FRITTERS

"This is my great-grandmother's recipe," says Lucille A. Grow, of the North Creek Valley Grange in Washington State.

MAKES 4 TO 6 SERVINGS

4 medium-sized ears fresh corn	2 eggs, lightly beaten
2 tablespoons flour	Vegetable cooking spray, butter or oil for the griddle
½ teaspoon baking powder	
½ teaspoon salt	Butter for the table
½ teaspoon granulated sugar	

1. Over a large bowl, cut the kernels from the cobs. Scrape the cobs well. Blend in the flour, baking powder, salt, sugar and eggs.
2. Grease a griddle and heat it to moderately hot (a drop of water flicked onto the grill should bounce). Pour the batter by the half-cupful onto the griddle. Do not crowd. Cook until brown on both sides. Serve with butter.

MAKE SOME CHANGE, THANK THE GRANGE

The yeoman farmer, 1873.

"I work in the Grange booth at the Tulelake-Butte Valley Fair, in northern California," says Mary Victorine. She and her husband Joe are retired potato farmers. ("We raised them till the kids left home. Two don't raise potatoes—you need a crew.")

"The Tulelake Grange started here in 1932. We're very active, and we have about a hundred families. As you know, the Grange is effective legislatively. We started the national Bottle Bill movement, right here in Tulelake! We have a lot of wind, and the cans that people threw along the highway were blowing into the fields, clogging the farm machinery. Grange member Paul Tschirky wrote a resolution about redeeming cans for money, to stop the mess. The resolution got passed along through the various Grange levels, all the way to Washington. The Bottle Bill passed. It's a lot cleaner around here now, except that Oregon gives five cents for a can, and California only two cents. We're on the border here, and I see scavengers who'll pick up the Oregon cans and throw the California ones back."

INDEX

PICTURE CREDITS

Page numberts are in **bold face.**

CHAPTER 1: **3,** Euclid Farnham; **6,** Euclid Farnham; **7** Champlain Valley Exposition; **10,** Hopkinton State Fair; **12,** Barbara Corson; **13,** Plymouth State Fair; **16** (*both*), Deerfield Fair; **18,** Northern Maine Fair; **19,** Gaylen Flewelling; **20,** Fryeburg Fair; **21,** Beverly Hegmann; **22,** Beverly Hegmann; **23,** Topsfield Fair; **27** (*left*), Connie Marra; (*right*), Fiesta Shows; **28,** The Rhode Island State Grange; **30,** Goshen Fair; **32** (*all*), Eastern States Exposition; **33,** Eastern States Exposition.

CHAPTER 2: **36,** Long Island Agricultural Society; **37,** Beverly Hegmann; **40,** Wickham family; **43,** Fred Briggs collection; **48,** Russell family; **50,** Grandma Moses Properties Co., New York, © 1987; **53,** Erie County Fair & Expo; **56,** Allan M. Herdman; **59,** Flemington Agricultural Fair; **62,** Bloomsburg Fair; **63,** Beverly Gruber; **65,** Historical society of Berks County; **68,** Hookstown Fair; **71,** Beverly Gruber; **72,** C. A. Muraski; **74** (*both*), Delaware Agricultural Museum & Village; **76,** Delaware State Fair; **78,** Kent County Fair; **81,** Morris family; **82,** Montgomery County Historical Society; **84,** Montgomery County Fair; **85,** Rockingham County Fair; **86,** Pat Murphy, *Harrisburg Daily News-Record*; **87** (*both*), Jackson County Junior Fair; **88,** B. Waskey.

CHAPTER 3: **90,** Cabarrus County Fair; **93,** Pendleton District Commissioner; **97,** John Bromel, *The Lebanon Enterprise*; **99,** John Bromel, *The Lebanon Enterprise*; **101,** *The Lebanon Enterprise*; **103,** Shelby County Fair & Horse Show; **108,** Appalachian Fair; **110,** Wilson County Fair; **112,** Georgia Mountain Fair; **115,** Lake County Chamber of Commerce; **116,** Citrus County Fair; **118,** James E. Strates Shows, Inc.; **120,** James E. Strates Shows, Inc.; **121,** Dade County Youth Fair & Expo; **122,** Dade County Youth Fair & Expo; **124** (*all*), Creative Putlet; **127,** Boyd family; **130,** *The Carthaginian.*

CHAPTER 4: **134** (*top*), Bercy La Fleur; (*bottom*), Calcasieu-Cameron Fair; **136,** Calcasieu-Cameron Fair; **137,** Washington Parish Free Fair; **139,** Washington County Fair; **141,** Garland County Fair; **142** (*all*), Garland County Fair; **145,** Will Rogers Memorial, Claremore, OK; **149,** Payne County Fair; **151,** Red River Valley Fair; **155** (*both*), State Fair of Texas; **158,** Blanco County Fair Association; **159,** Ava Johnson Cox; **160,** Crystal Sultemeier; **161,** Panhandle South Plains Fair.

CHAPTER 5: **164** (all), Ohio State Fair; **165,** Katherine Grimm; **166,** Canfield County Fair; **168,** Clark County Fair; **170,** Dorothy Drake; **171,** National 4-H Council Resource Center; **173,** Delaware County Fair; **176,** Alan Snead; **177,** Elkhart County Fair; **180,** Conn Foster Museum, Travers City, MI; **181,** Northwestern Michigan Fair; **184,** Jeff Lee, *Press Democrat*; **187,** Upper Peninsula State Fair; **190,** Sheboygan County Fair Assoc.; **191,** Sheboygan County Fair Assoc.; **192,** Shawano County Fair; **192,** *Shawano Leader*; **195,** *Peoria Journal-Star*; **196,** Virginia Williams; **197** (both), The Sandwich Fair.

CHAPTER 6: **200,** American Dairy Association; **201,** M. A. Gedney Co.; **203,** Joyce Agnew; **204,** Minnesota State Fair; **206,** Olmsted County Historical Society Archives; **208,** Pembina County Fair; **209,** Mr. and Mrs. Marvin Briese; **211,** Sioux Empire Fair; **213,** Midwest Old Threshers; **217,** Midwest Old Threshers; **218,** Madison County Sheep Producers; **220,** Buchanan County Fair; **223,** *The Phonograph Herald*; **225,** National 4-H Council Resource Center; **227,** Nebraska State Fair; **228** (*both*), Nebraska State Fair; **229,** *The News-Leader*; **231,** Ozark Empire Fair; **232,** Ford County Fair; **233,** Garrett McClure; **234,** Ford County Fair; **235,** Rooks County Free Fair.

CHAPTER 7: **238,** Lea County Fair & Rodeo; **239,** Stacy Medlin Reid; **240,** *Hobbs Daily News-Sun*; **245** (*left*), Strack family; (*right*), Darren Houn, *Reporter-Herald*; **247,** Larimer County Fair; **248,** BaxterBlack; **251,** Sweetwater County Fair; **252,** Sweetwater County Fair; **255,** KPRK, Livingston; **260,** Twin Falls County Fair & Rodeo; **262** (*both*), Rooster Kersten; **263,** Rooster Kersten; **264,** Calgary Stampede; **268,** "Sam" De Leeuw; **270,** Maricopa County Fair, Inc.; **273,** Yavapai County Fair; **275,** Canfield Fair; **276,** Eureka County Fair.

CHAPTER 8: **280,** National Date Festival; **282,** National Date Festival; **283,** Orange County Fair; **287,** The Big Fresno Fair; **292,** Calaveras County Fair; **296,** *Press-Democrat*, Santa Rosa; **298,** Tulelake-Butte Valley Fair; **304,** Clackamas County Fair; **307,** Tillamook County Fair; **310,** Tillamook County Fair; **311,** Denise Dykstra; **313,** *Grand View Herald*; **314,** Qualex, Inc.; **316,** PCI Inc.; **318,** N.E.W.S. Photo; **320,** Alaska State Fair; **323,** Matt Thayer, *The Maui News*; **324,** Maui County Fair; **327,** Toshio Endo.

CHAPTER 9: **329,** National 4-H Council Resource Center; **330** (*top*), United States Department of Agriculture Extension Service; (*bottom*) National 4-H Council Resource Center; **332** (*both*), Quarty family; **334,** Hurld family; **336** (*both*), National FFA Organization; **338,** National FFA Organization; **339** (both), The National Grange of the Order of Patrons of Husbandry; **342,** The National Grange of the Order of Patrons of